SOIL
EROSION
RESEARCH
METHODS

D1241968

R. Lal, Editor

SOIL EROSION RESEARCH METHODS

SOIL
AND WATER
CONSERVATION
SOCIETY

Soil and Water Conservation Society
7515 Northeast Ankeny Road
Ankeny, Iowa 50021-9764

Subcommission C: Soil Conservation and Environment
International Society of Soil Science
P.O. Box 353
6700 AJ Wageningen, The Netherlands

Copyright © 1988 by the Soil and Water Conservation Society
All rights reserved
Manufactured in the United States of America

Library of Congress Catalog Card No. 88-11381

ISBN 0-935734-18-X

$16.00

Library of Congress Cataloging-in-Publication Data

Main entry under title:

Soil Erosion Research Methods / edited by R. Lal
   p.   cm.
   Includes index.
   ISBN 0-935734-18-X
   1. Soil erosion—Research—Methodology. 2. Soil
conservation—Research-Methodology. I. Lal, R. II. Soil and
Water Conservation Society (Ankeny, Iowa) III. International
Society of Soil Science. Subcommission for Soil Conservation and
Environment.
S626.S65 1988
631.4'5'072--dc19   88-11381
                                                                CIP

Contributors

Samir A. El-Swaify
Professor, Department of Agronomy and Soil Science, College of Tropical Agriculture and Human Resources, University of Hawaii at Manoa, Honolulu

Klaus W. Flach
Special Assistant for Science and Technology, Soil Conservation Service, U.S. Department of Agriculture, Washington, D.C.

G. R. Foster
Head, Department of Agricultural Engineering, University of Minnesota, St. Paul

R. Lal
Associate Professor, Department of Agronomy, Ohio State University, Columbus

K. C. McGregor
Agricultural Engineer, Sedimentation Laboratory, Agricultural Research Service, U.S. Department of Agriculture, Oxford, Mississippi

L. D. Meyer
Supervising Agricultural Engineer, Sedimentation Laboratory, Agricultural Research Service, U.S. Department of Agriculture, Oxford, Mississippi

C. E. Murphree
Agricultural Engineer, Sedimentation Laboratory, Agricultural Research Service, U.S. Department of Agriculture, Oxford, Mississippi

C. K. Mutchler
Supervisory Research Hydraulic Engineer, Sedimentation Laboratory, Agricultural Research Service, U.S. Department of Agriculture, Oxford, Mississippi

C. W. Rose
Foundation Professor, School of Australian Environmental Studies, Griffith University, Brisbane, Queensland

E. L. Skidmore
Soil Scientist, Northern Plains Area, Agricultural Research Service, U.S. Department of Agriculture, Manhattan, Kansas

M. A. Stocking
Consultant, Food and Agriculture Organization, United Nations, and Soil Scientist, Overseas Development Group, University of East Anglia, Norwich, England

D. E. Walling
Hydrologist, Department of Geography, University of Exeter, Exeter, England

Contents

Contributors,   v

Foreword,   ix

Preface,   xi

1 Soil erosion by wind and water: Problems
and prospects   1
   R. Lal

2 Laboratory and field plots for soil erosion studies   9
   *C. K. Mutchler, C. E. Murphree,
   and K. C. McGregor*

3 Measuring sediment yield from river basins   39
   D. E. Walling

4 Rainfall simulators for soil conservation research   75
   L. D. Meyer

5 Modeling soil erosion and sediment yield   97
   G. R. Foster

6 Research progress on soil erosion processes and
a basis for soil conservation practices   119
   C. W. Rose

7 Erodibility and erosivity   141
   R. Lal

8 Assessing vegetative cover and management
   effects   163
   M. A. Stocking

9 Monitoring soil erosion's impact on crop
   productivity   187
   R. Lal

10 Wind erosion   203
   E. L. Skidmore

   Appendix,   235

   Index,   237

Foreword

The international community is becoming increasingly aware of the dangers that soil erosion and sedimentation pose to sustainable agriculture and the overall stability and quality of the environment. Formation of Subcommission C: Soil Conservation and Environment by the International Society of Soil Science not only recognizes these dangers but also reflects a commitment by all soil scientists to use their skills in developing effective solutions.

This publication is one of the first major activities of this subcommission. Members of the subcommission and other interested scientists perceived the need for a volume that would provide an overview of research methods to assess the magnitude and impact of soil erosion. They believed that such a publication would help disseminate proven technology while also stimulating research on better methods of assessing the damages of erosion throughout the world and of developing and applying soil-saving farm practices.

We are grateful to the authors for their contributions, to the institutions for which they work for providing the opportunity for research, and to the Soil and Water Conservation Society for its assistance in publishing this book at a price that will make it accessible to scientists in all countries. We also acknowledge the financial support from the International Institute for Tropical Agriculture and the U.S. Agency for International Development, without which this publication would not have been possible.

Samir A. El-Swaify, Chairman,
    and Klaus W. Flach, Past Chairman
Subcommission C: Soil Conservation and Environment
International Society of Soil Science

Preface

Soil erosion is a major environmental concern of modern time. As an emotional and debatable issue, two schools of thought have developed about the consequences of soil erosion.

One school, comprised of ecologists and environmentalists, believes that accelerated erosion is a cancer on the land that rapidly is depleting the soil's productive capacity and causing the pollution and eutrophication of natural waters. In support of their claim, adherents of this school cite the fact that the annual loss of agricultural land due to soil erosion and desertification is as high as 3 million hectares and 2 million hectares, respectively. At this high rate of agricultural land loss, they fear that the world may lose from one-fifth to one-third of its agricultural land by the year 2000. The annual global discharge of sediment into the oceans has increased, presumably, from 2 billion tons before the development of settled agriculture to 24 billion tons now.

The other school believes that the effects of soil erosion are drastically exaggerated, though they accept that excessive soil erosion can be a problem. Some soil scientists sincerely believe that the loss of topsoil can be compensated for easily by adding a few kilograms of additional nitrogen to the soil. They further argue that on-site crop yield reductions may be compensated for by increased yields at the site of deposition. The development of most fertile valleys, afterall, they contend, is attributed to erosion in the upper reaches of the catchment. The proponents of this school believe that the "crying wolf attitude" can be counterproductive.

The effects of soil erosion on this resource specifically and the environment generally are easily exaggerated when factual information is scarce. Despite the voluminous literature on the global, regional, and national problems of soil erosion, quantitative and reliable data on the magnitude of problem are indeed scarce. Furthermore, there are few if any checks to verify the validity of available statistics on the magnitude of soil erosion. Most available information, especially that from the tropics, is based on reconnaissance surveys or on experiments that lack a standardized methodology. Such an information base may be of some use in creating public awareness, but it is of little value in developing and implementing strategies to prevent or control erosion. Whatever quantitative data are available, particularly from the tropical countries, usually are obtained with unstandardized methodology. The size of runoff plots, the method of collecting runoff and sediment samples, use of a partitioning or multislot system, design of storage tanks, and other factors differ among researchers. Equipment design and the size of the field plots are dictated more by budgets and the available manpower than by the need for scientific precision and data accuracy. A survey of the equipment and methodologies used in field plot experiments would indicate a wide range of techniques used in evaluating soil erosion. The results thus obtained are methodology-specific and cannot be compared among widely scattered experimental sites.

The belief that "some data are better than no data" has produced erroneous results, which have led to costly mistakes in designing ineffective erosion control measures. Erosion research, although capital- and labor-intensive, is not expensive. There is indeed a paucity of results produced from well-designed and properly equipped erosion experiments. It is precisely that data from such experiments that scientists and policymakers require for developing appropriate resource management strategies.

It is for these reasons that the newly formed Subcommission C: Soil Conservation and Environment of the International Society of Soil Science undertook this project of standardizing erosion research methodology. While preparing the outline, the subcommission carefully chose the topics to reflect the most basic of research needs. In addition to professional competence, authors were selected on the basis of their experiences in diverse ecological and geographical regions.

The book includes 10 chapters. The first addresses the issues of evaluating soil erosion problems with unstandardized methodologies and of data precision and reliability. Chapters 2 through 9 deal with

methodologies involved in the laboratory, field runoff plots, and large
river basins; the design and use of rainfall simulators; modeling soil
erosion processes; methods of monitoring erodibility and erosivity
and of the canopy cover; and assessing the impact of erosion on productivity. The last chapter discusses the important topic of wind erosion and available techniques to measure and predict its magnitude.

The project, initially planned and discussed in January 1983, got
off to an excellent start. All authors responded enthusiastically and
prepared and revised their manuscripts on schedule. I express my
profound appreciation to each of them for their sincere efforts to
prepare top quality manuscripts on schedule. Once the manuscript
was assembled, the project ran into unforeseen financial difficulties.
Once again, I express my thanks to all authors for their cooperation
and understanding of this difficult situation. Sincere efforts to pro-
cure the financial support were made by Dr. Klaus Flach, president
of Subcommission C; by Dr. W. Sombroek, secretary general of the
International Society of Soil Science; and by Dr. Ray Meyer and
Dr. Hari Eswaran of the U.S. Agency for International Development. Dr. Flach left no stone unturned in getting this project completed. The financial support received from the International Institute of Tropical Agriculture and from USAID made it possible for
the Soil and Water Conservation Society to publish this important
document.

This publication is an important step toward realizing the dream
of standardizing erosion research methodologies. It has been an honor
to work with members and officials of Subcommission C and many
accomplished and distinguished scientists who prepared their respective manuscripts meticulously and on schedule. Finally, I received
full cooperation and support from Messrs. Max Schnepf and James
Sanders of the Soil and Water Conservation Society. It has been a
pleasure working with them and with other members of their editorial
staff. I hope that this association will help the scientific community
in moving a step closer to understanding and solving the severe problems of soil erosion.

Rattan Lal, Editor

January 1988

1

R. Lal

Soil erosion by wind and water: Problems and prospects

That accelerated soil erosion is a serious global problem is widely recognized. What is difficult to assess reliably and precisely, however, are the dimensions—the extent, magnitude, and rate—of soil erosion and its economic and environmental consequences. Information readily available in the literature often is based on reconnaissance surveys and extrapolations based on sketchy data.

Judson (8) estimated that river-borne sediments carried into the oceans increased from 10 billion tons a year before the introduction of intensive agriculture, grazing, and other activities to between 25 billion and 50 billion tons thereafter (1). Dudal (4) reported that the current rate of agricultural land degradation worldwide by soil erosion and other factors is leading to an irreversible loss in productivity on about 6 million ha of fertile land a year. According to the United Nations Environmental Program (15), crop productivity on about 20 million ha each year is reduced to zero or becomes uneconomic because of soil erosion and erosion-induced degradation. On the basis of information available worldwide on soil erosion, Higgins and associates (7) reported that crop yields in rainfed areas might decrease 29 percent over the next 25 years. Since the beginning of settled agriculture, soil erosion has destroyed about 430 million ha of productive land (9). Buringh (2) estimated that the annual global loss of agricultural land is 3 million ha due to soil erosion and 2 million ha due to desertification. Of the total annual sediment load of 1 billion tons carried by rivers from the continental United States, about 60 percent is estimated to be from agricultural land (10, 11). The off-site damages caused by sediment in the United States are estimated at $6 billion annually (3), including $570 million for dredging several

1

million cubic meters of sediment from U.S. rivers, harbors, and reservoirs (12).

Equally ruinous are the adverse effects of wind erosion. Although wind erosion is less than water erosion on a world scale, the problem is severe in many semiarid and arid regions. However, even less research data is available for wind erosion than for water erosion. The basic principles governing wind erosion processes and erosion control are similar to those for water erosion. Nonetheless, the specific cause-and-effect relationships, the magnitude of wind erosion in different ecologies, and the effectiveness of erosion control practices on management systems have not been investigated widely.

The need for improved data

These frightening statistics leave many questions unanswered on data sources, methods of data collection and extrapolation, and data accuracy and reliability. Soil erosion research is a capital-intensive, time-consuming exercise. Global extrapolation on the basis of few data collected by diverse and unstandardized methods can lead to gross errors. Erroneous and unreliable information is worse than having no information because it can lead to costly mistakes and misjudgements on critical policy issues.

There is an urgent need for standardizing methodologies to increase the reliability and accuracy of data on soil erosion. The information should be factual and devoid of emotional rhetoric. Some basic problems with data collection are as follows:

Data extrapolation. Extrapolations based on limited data and to regions outside the ecological limits in which experiments were conducted are often misleading. Two examples are the attempts to compile suspended sediment yield maps for Africa and South America (Figure 1). Fournier (6) determined sediment yield in Africa south of the Sahara Desert on the basis of extrapolation of relationships between specific sediment yield and catchment relief in other regions, for example, continental Europe. In another case, Strakhov (13) used meager data and extrapolated it to the entire African continent. These two maps differ both in trend and in sediment yields by several orders of magnitude.

More recently, Walling (16), in collaboration with UNESCO, compiled the most up-to-date map based on sediment yield data for Africa

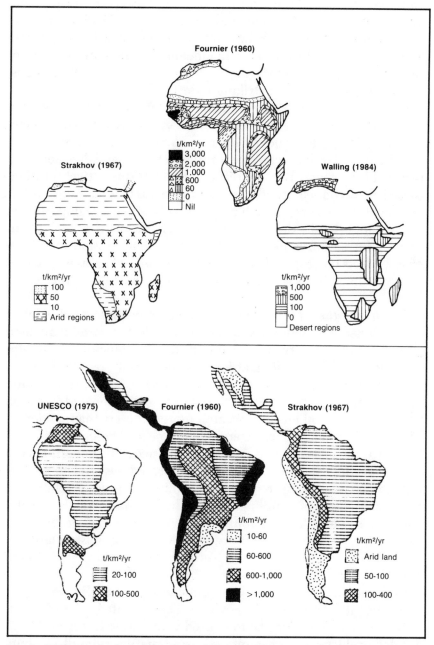

Figure 1. Comparison of suspended sediment yield maps of Africa (top) and South America (bottom) prepared by Fournier (6), and Strakhov (13), UNESCO (14) and Walling (16).

and other regions of the world. This map, based on existing data, differs from both the Fournier and Strakhov maps. Walling indicates the dangers of insufficient data and points out that his map is still tentative, subject to further refinements. The magnitude of error caused by using Strakhov's map or Fournier's map, if Walling's is, in fact, more reliable, is striking indeed.

Rather than quantitatively measuring soil loss, field assessment often involves a qualitative evaluation of soil erosion. Soil movement is visually judged to be none, slight, moderate, or severe. Assessment of soil erosion by this technique is rather selective and depends upon personal judgement. It is also difficult to relate quantitively erosion-caused effects on soil properties and crop yields to soil erosion. Information of this nature, though useful, is only of relative importance. Unless supported by data on soil profile characteristics, such qualitative information is hard to generalize. Slight erosion may have severe effects on crop yields in a marginal soil, whereas severe erosion may have little effect on crop yields in a deep, fertile soil.

Standardization of methodology and data reliability. A wide range of plot dimensions and runoff and erosion collection systems are used to monitor the severity of erosion in different ecologies. The equipment used is often insufficient for expected runoff; overflow thus is a common problem. A range of multidivisor systems, storage and siltation tanks and reservoirs, and sediment collection and filtration methods makes soil erosion research more of an art than a science.

In collecting samples for determining suspended sediment and bed-load measurements in rivers, Walling (*16*) listed some important considerations, for example, equipment and procedures used for collecting samples, sampling frequency, sampling location in relation to river bank, depth of sampling, and methods of data calculation and analysis. In fact, data reliability is one of the serious problems in soil erosion research. Erosion rates assessed by an unstandardized methodology are unreliable. Regrettably, the literature is polluted with such data.

Attempts have been made to estimate soil erodibility from empirical equations developed elsewhere. The estimated values of erodibility often differ from the measured values by a factor of 2 to 5. Similar problems are encountered in estimating rainfall erosivity without prior knowledge of drop size distribution, intensity, and kinetic energy and the relationships among these variables. Isoerodent maps prepared on the basis of guestimated R factors can do more harm than good.

Record duration and data continuity. Data should be obtained with standard equipment and for a reasonable length of time to provide long-term trends. Soil erodibility and detachability are time-dependent functions. Rain and wind erosivity assessments should be made based on long-term trends in rainfall or wind velocity. The sediment load in rivers is subject to seasonal and annual fluctuations. The interannual variability of these parameters can only be ascertained from long-term data obtained using standard methods. Spot measurements of erodibility, erosivity, or load can be misleading.

Toward standardization

Soil erosion is among the most critical environmental hazards of modern times. Vast areas of land now being cultivated may be rendered unproductive or at least economically unproductive if erosion continues unabated. Although accelerated erosion and erosion-caused soil degradation are unmitigated evils, quantitative data on the rate of soil erosion for different environments and land uses are not available in many regions. Guesses, emotional statements, and qualitative assessments should be replaced by facts and figures supported by verifiable data. There is a need for more basic research to generate data that is accurate, reliable, and obtained by standardized methods. There is much talk about the severity of erosion, yet little is known about quantitative rates of soil erosion. What are the criteria to evaluate the degree of soil degradation caused by erosion? How can one assess the impact of soil erosion in economic terms?

For results to be comparable and for their extrapolation from one region to another, it is important that methodologies be standardized. This volume is an attempt to define standard methods and techniques used for soil erosion and sedimentation research. The emphasis is on basic data collection. Existing models also are explained, and those that can be used as tools to identify the knowledge gaps and to extrapolate the data to other regions are discussed.

REFERENCES

1. Brown, L. R., and E. C. Wolf. 1984. *Soil erosion: Quiet crisis in the world economy.* Paper No. 60. Worldwatch Institute, Washington, D.C. 50 pp.

2. Buringh, P. 1981. *An assessment of losses and degradation of productive agricultural land in the world.* Working Group on Soils Policy, Food and Agriculture Organization, United Nations, Rome, Italy.
3. Clark, E. H. 1985. *The off-site costs of soil erosion.* Journal of Soil and Water Conservation 40: 19-22.
4. Dudal, R. 1981. *An evaluation of conservation needs.* In R.P.C. Morgan [editor] *Soil Conservation, Problems and Prospects.* J. Wiley & Sons, Chichester, England. pp. 3-12.
5. Fournier, F. 1960. *Climat et erosion.* Presses Universitaires de France, Paris.
6. Fournier, F. 1962. *Map of erosion danger in Africa south of the Sahara.* Commission for Technical Cooperation in Africa, European Economic Community, Paris, France.
7. Higgins, G. M., A. H. Kassam, and L. Naiken. 1982. *Potential population supporting capacities of land in developing world.* Project INT/75/P13. Food and Agriculture Organization, United Nations. Rome, Italy.
8. Judson, S. 1981. *What's happening to our continents.* In B. J. Skinner [editor] *Use and Misuse of Earth's Surface.* William Kaufman, Inc., Los Altos, California. pp. 12-139.
9. Kovda, V. A. 1983. *Loss of productive land due to salinization.* Ambio 12: 91-93.
10. Larson, W. E., F. J. Pierce, and R. H. Dowdy. 1983. *The threat of soil erosion to long-term crop production.* Science 219: 458-465.
11. National Academy of Science. 1974. *Productive agriculture and a quality environment.* Washington, D.C. 189 pp.
12. Pimental, D., J. Allen, A. Beers, L. Guirand, R. Linder, P. McLaughlin, B. Meer, D. Mussonds, D. Perdue, S. Poisson, S. Siebert, K. Stoner, R. Salazar, and A. Hawkins. 1987. *World agriculture and soil erosion: Erosion threatens world food production.* BioScience 37(4): 277-283.
13. Strakhov, N. M. 1967. *Principles of lithogenesis* (volume 1). Oliver & Boyd, Edinburgh, England.
14. United Nations Educational, Scientific and Cultural Organization and International Association of Hydrological Sciences. 1975. *Gross sediment transport into the oceans.* Paris, France.
15. United Nations Environmental Program. 1980. *Annual Review.* Nairobi, Kenya.
16. Walling, D. E. 1984. *The sediment yields of African rivers.* Publication 144. International Association of Hydrological Sciences, Wallingford, England. pp. 265-283.

2

C. K. Mutchler, C. E. Murphree, and K. C. McGregor

Laboratory and field plots for soil erosion studies

Soil erosion research must be based on experimental results of some form. Often, laboratory and field plots are used to obtain experimental data for predicting and evaluating soil erosion and sediment yield.

Laboratory and field plots are important only as tools to acquire data to accomplish some research objective. Plots must be designed to furnish data that can be analyzed to test hypotheses and to provide some knowledge or technique useful for soil conservation.

A project outline

The first step in any research project should be a working outline that gives specific instructions for performing and completing the experiment. First is a set of objectives and justification for doing the research. These elements are necessary for two reasons: to obtain money and materials for the research and to guide the research through the life of the project.

The next part of the outline is a literature review. This ensures that the research does not reinvent the wheel. The length and time spent on a literature review depends upon the experience of the researcher and the technical guidance provided.

The procedures part of the outline describes in detail the instrumentation required and data collection methods. No single part of the outline can stand alone, except perhaps the objectives and justification. The procedures section in particular must support all other parts of the research outline.

The last part of the outline is a plan for analysis. Experiments using erosion plots generally are expensive. Therefore, researchers must take care that all the needed data are collected to complete the analysis. Also, the plan for analysis helps to ensure that questions in the objectives will be answered.

Careful preparation of a research outline is critical because of the peculiar nature of erosion research. Most procedures tend to destroy the erosion plot. This is especially so in laboratory experiments that require expensive, time-consuming efforts to prepare a plot for application of a simulated rainstorm. In short, all data must be collected correctly the first time to prevent repreparation of the laboratory plot or a long wait to allow plot conditions to be established once again.

Physical modeling of the soil surface with or without plants must always be done in full scale. Sand, silt, and clay particles cannot be scaled for reduced-scale experiments. Agricultural soils have many chemical and physical properties, particularly those of cohesion, that differ for each soil type. Also, land invariably has some runoff channelization tendencies that are modified by cropping patterns. These factors require plots to be full size. Likewise, plants are difficult to use except in a full-size model. Most important, rainfall cannot be scaled. Every raindrop has its own shape and velocity, depending upon its size. Consequently, plot size is important to represent properly the soil and plant situation being sampled. The research outline should ensure correct use of the plot.

The erosion process

The term "erosion" often is used as an all-inclusive word to describe the wearing down of a landscape. However, we prefer to define erosion as the detachment or entrainment of soil particles, thus distinguishing it from deposition or sedimentation and sediment transport. Herein, we use the term sediment yield to refer to the amount of eroded material that passes a designated point at the outflow end of a plot, field, channel, or watershed (catchment). The term soil loss, expressed as a quantity per unit area and time, often is used for small plots. Even in the most erodible situations, soil loss or sediment yield is limited by the transport capacity of the runoff. As runoff flows through a watershed, changes in topography, vegetation, and soil characteristics often reduce this transport capacity. There are various opportunities for deposition as a result. Gross erosion in a watershed, as predicted from plot results, must be reduced by a sediment delivery ratio (unique for each watershed) to obtain an expected sediment yield for a watershed.

Basically, any thin depth of water on a slope concentrates as it flows downhill because of the surface tension properties of water. The result of this phenomenon can be seen after landscapes have eroded under the flow concentrations. Researchers have devised various systems of stream orders to study the results of runoff concentrations on the land. Although farm operators are more concerned with the farmable areas between stream channels, the same general effect of flow concentrations is observable in areas between channels that cannot be crossed in a farming operation.

Raindrop impact erodes the land surface between rills and also initiates transport of detached soil particles to the rills. The complete transport system involves the initial movement of soil from the point of raindrop impact to small rills, to larger rills, to ephemeral channels, and to continuously flowing rivers. Erosion and sedimentation can occur at any point in the system. This process is called geologic erosion when it occurs without human influence. Alternatively, the process is called accelerated erosion when activity by humans causes increased erosion, such as land disturbed for crop production.

During agricultural operations, cultivation, row direction in planting, and water management practices designed to convey runoff without allowing excessive erosion all change the natural rill system. In row-crop agriculture, rills generally form in tillage paths, especially where the crop is cultivated. For up-and-down-hill rows, the rills enlarge as runoff amounts and rill erosion increase. For contour rows, rill enlargement is slower because of lower runoff velocities. However, the capacity of rills parallel to the rows often is exceeded under large storms, and rills parallel with the slope are formed. Terraces built across a slope interrupt downslope rill patterns and keep rill formation at the elementary stage.

Knowledge of the erosion process is necessary because erosion studies use plots that are samples of this "big picture." To be valid, results of such research must be from a representative sample of the landscape. Researchers, of course, use various plot types to study soil erosion. The major concern is that any plot is a sample of the landscape, and it must furnish data that can be analyzed and formulated for use on the land.

In summary, then, a land area is comprised of rills and channels and the land between them. More specifically, farmland can be described as rill and interrill areas and the erosion processes as interrill erosion, rill transport, rill erosion, and deposition. Plots can represent full-size areas of interrill erosion or rills separately, or they may represent areas that include all of the processes. Thus, erosion plots fall into three categories: small plots, USLE (universal soil loss equation) plots, and large plots.

Small plots

Researchers use small plots to study separately rill or interrill erosion. Such plots are typified by the square-meter plots used by Meyer and Harmon (12) to study interrill erosion on cropland in row crops. Laboratory plots, used to isolate some small part of the erosion process for intensive study, are also common because of space limitations (Figure 1).

Figure 1. A laboratory erosion plot used to study the effects of crop residue placement on soil erosion. The 46-cm × 46-cm plot is surrounded by a border 23 cm wide. Runoff enters a slot at the bottom of the small plot and is collected in a container for measurement and sediment analysis. Rainfall is applied with a nozzle-type rainfall simulator mounted 3 m over the plot.

Small plots usually use rainfall simulation (see chapter 4).

There is extensive literature in soil physics on the use of small soil plots. Here, we discuss only those principles that are important to erosion research.

Objectives. Generally, researchers use small plots to study basic erosion phases that are difficult to study in detail on larger plots, for example, surface sealing, aggregate stability, raindrop detachment, and splash transport. Note that these topics deal with processes in the interrill area. Experiments with rill processes require plots long enough to develop sufficient runoff. Meyer's studies of rill and interrill erosion are examples of small plots used in the field under simulated rainfall.

Small plot experiments should have objectives complementary to field plot experimentation or application. Moldenhauer, for example, used small plots to study the effects of soil characteristics on erodibility (13). He used laboratory plots to isolate certain phases of the erosion process, then examined and verified the results with rainfall, climate, and soils in the field.

A currently popular objective is to develop a mathematical model of the erosion process. Small plots then are used to develop or verify basic operating equations that govern the physical process of soil erosion.

Justification. The greatest justification for small plots is their utility in studying the basic aspects of soil erosion in detail. A researcher can control the parameters of the erosion process more closely on small plots than on large plots. Experiments using small plots are more justifiable if the anticipated results have a planned application. In most cases, these experiments provide basic concepts and knowledge required for efficient developmental research.

Procedures. Problems that influence study procedures on small plots include those related to size and duplication of the soil as it is found in a field.

Edge effects, which must be considered in all plots, are magnified with small plots. Edge effects include the movement of splash into or out of a plot, the addition of eroded soil to a plot treatment from an erodible soil ridge, a barrier effect on soil water and plant root movement, and other boundary effects caused by the restriction of splash or runoff on a plot. Laboratory plots also have a boundary under the soil plot.

Edge effects can be minimized by collecting all or almost all of the splash, suitable generally for single-drop experiments using a soil target as exemplified by Al-Durrah and Bradford (1) or by using a border of thin metal, for example, to delineate the plot and maintaining a border area around the plot in the same condition as the plot area. For very small plots in the field or laboratory, a 1-m border is desirable to ensure that as much splash will enter the plot as leaves the plot. Splash from raindrop impact is greatest from a surface covered with a very thin layer of water. Thus, results from any small plot will be biased if the surface water is not maintained in a condition representative of the treatment in the field. This thin layer is best described as the amount of water on an interrill surface when runoff is initiated. The phenomenon is demonstrated by examining records of runoff and sediment concentration from a plot under simulated rainfall; invariably, the concentration values at the beginning of runoff are higher than at any other time in the test.

Small laboratory plots must have a bottom to hold the soil. The design of a bottom depends, of course, upon the objective of the experiment. For erosion experiments where runoff and soil loss are the primary indicators of differences in the treatments, an open bottom made of screens overlain by cloth often is used to allow free passage of soil water. This is representative of an extreme condition of layered soil, such as loam underlain with a much coarser material. Moldenhauer (13) found that using suction under a plot caused considerable variation, especially with thin layers of soil. Consequently, he used a 15-cm-deep soil layer with free drainage. Use of such a system is supported by considering the events that occur during a simulated rainfall experiment. Most simulator storms are 30 to 60 minutes long, and surface saturation must occur before runoff begins. Hydrographs from field plots often show a constant runoff rate (and constant infiltration rate) after 45 to 50 minutes for initial dry runs and after 5 to 15 minutes for wet runs made 4 to 24 hours after the dry run. The point is that the time required for infiltration and saturation of a 15-cm soil profile is longer than the time used to evaluate erosivity with a rainfall simulator at 6.4 cm/hour. Of course, lower rainfall rates change these times, and there are some unique volcanic soils with extremely high infiltration rates.

Another problem in simulating a field soil occurs because soil in the laboratory plot has been disturbed. One solution is to use soil only from the plow layer in the field, air dry it for storage, sieve it through a large screen before placing it in the plot container, and pack it into the plot container in layers (13). All soil must be prepared uniformly to minimize differences between replicates.

Many problems with plot preparation can be avoided by covering field locations to allow only simulated rainfall to fall on the plots. Thus, field cultivation, rainfall compaction, and other events need not be simulated.

Analysis. It is difficult to discuss particular analyses of data from small plots because they are used for a variety of objectives. The same may be said of the objectives and justification statements discussed earlier. Most important, researchers must consider analysis when planning the experiment as well as use of the results either in field research or in application of the research by conservationists or landowners.

USLE plots

The second type of erosion research plot is large enough to represent the complete process of rill and interrill erosion (Figure 2). A good example are plots used to develop the USLE (27). These plots measure erosion from the top of a slope where runoff begins. They are sufficiently wide to minimize edge or border effects and large enough for development of downslope rills. The plots are limited to rows parallel to the slope. The topographical parameters of slope length and steepness, as used in the USLE, are determined from plots of different lengths and steepnesses. Slope irregularity effects are studied with concave and convex plots. Surface roughness, residue cover, plant canopy, antecedent cropping effects, and other cropping and management effects are evaluated using specific crops and tillage sequences on standard plots. Predicting soil loss from other locations is made possible by an erodibility evaluation of each soil type and a rainfall erosivity estimate. To expand the use of data collected to field-size areas, a factor is used to reduce the erosion estimate for contouring, stripcropping, or terracing.

The USLE plot exemplifies a representative slope length. The USLE (27) is $A = RKLSCP$, where A is soil loss per unit area, R is the rainfall factor, K is the soil erodibility factor, L and S are the

Figure 2. The standard USLE plot is 22.1 m long and, in this case, 4 m wide (4 crop row widths). The plot is bordered with sheet steel strips driven into the ground and an endplate at the bottom end of plot that also holds a runoff collecting trough. Runoff is measured with an H-flume and water-stage recorder. An aliquot sample of storm runoff is taken with a Coshocton wheel sampler.

slope length and steepness factors, C is the cover and management factor, and P is the support practice factor. Only A, R, and K have dimensions. In customary English units, A is in tons per acre per year, R is in hundreds of foot-ton-inches per acre-hour-year, and K is in ton-acre-hours per hundreds of foot-ton-inch-acres. In SI units[1], A is in t/ha·y, R is in MJ·mm/(ha·h·y), and K is in t·ha·h/(ha·MJ·mm), where t is metric tons. Note that R customarily represents the average annual accumulated storm erosivity index (EI). For a cropstage, the EI terminology is retained. For a cropstage, the soil loss equation is A (cropstage) = (EI)KLS(SLR)P, with the same time unit used for A and EI. Also, SLR (soil loss ratio) is used instead of C for time periods not equal to 1 year. Referring to data from a plot, R is the accumulated rainfall EI that produced the soil

[1]The USLE was developed using U.S. customary units. Conversions of units to the International System of Units are used according to Foster and associates (5). U.S. customary units are given in parentheses where it is thought helpful to avoid confusion.

loss, A, on a 9 percent slope, 22.1 m (72.6 feet) long, in continuous cultivated fallow. K, then, is the ratio of A/R. In practice, slopes of 9 percent for all soil types are not found, so the S factor is used to adjust the soil loss value to that expected from a 9 percent slope. Likewise, the L factor is used to adjust the soil loss to that from a 22.1-m length if some other slope length is used.

Objectives. The general objective for field plots in USLE-related research was to develop a soil loss prediction equation for conservation planning, which requires an average annual soil loss estimate. This is what the unmodified USLE furnishes. The equation, like the plots used for gathering its data base, is suitable only for estimating erosion along an up-and-down-hill profile; that profile must stretch from the top of a hill to the bottom, until the slope steepness decreases to the point that deposition occurs or a drainageway is encountered. Conservationists use the equation to design conservation plans on a worst-case basis. For example, the most erodible portion of a field in question is either cut out of the tillable area, used to control the farming practices in the entire field, or used as the basis for a compromise plan that may include such practices as terraces.

Justification. The major reason for early research leading to development of the USLE was to give working soil conservationists a means of quantifying soil losses on land treated with conservation practices in comparison with untreated land. Landowners are more easily persuaded to use soil conservation methods if, for example, a scientific estimate shows soil loss of 40 t/ha·y, which exceeds an allowable 11 t/ha·y, rather than saying that terraces are "good." Also, the USLE assures landowners that their respective fields are being evaluated on the same basis. Soil conservation promotion is difficult because significant soil loss rates of 20 to 40 t/ha·y are not noticed easily, even by trained observers. Gross mismanagement of land that causes a soil loss of 80 t/ha·y or more almost invariably results in large (10 to 12 cm deep) cropland rills and badly eroded drainageways. Erosion from channels cannot be estimated with the USLE, but channels can readily be observed in the field.

Recently, researchers have tried to modify the USLE and thereby extend its use beyond the original intent of providing an average annual soil loss estimate. Williams (23) developed the MUSLE (modified USLE), which uses procedures to obtain an average K, L, S, and C for small agricultural watersheds. He replaced the R factor with a runoff factor to give a storm estimate of sediment yield. However, this requires use of a storm estimate of runoff volume and peak runoff rate from the watershed.

Researchers also have employed the USLE to estimate sediment yield; the procedure uses a sediment delivery ratio (SDR) to account for deposition of eroded material as estimated by the USLE. The dissimilarity of watersheds has made formulation of sediment delivery ratios difficult.

The CREAMS model, designed to estimate storm sediment yield from small agricultural watersheds (22), also uses USLE factors. This model greatly extends the USLE by considering runoff, rill and inter-rill erosion, sediment sizes and densities, erosion of channels, and deposition of detached soil in parts of the watershed, including areas of ponded water. The justification for this model, in addition to that for the USLE, is to provide a storm model of sediment yield from the field or small agricultural watershed.

Researchers have made many other uses of USLE parameters to devise estimates of soil loss for purposes other than agriculture. A notable example is the adaptation of the USLE for use on forest land (3).

Procedures. As explained earlier, the USLE-type plot samples the combined processes of rill and interrill erosion. Experience has shown that 5 m is about the minimum slope length that will represent adequately a rill system in an up-and-down-hill plot. A better minimum length is 9 to 10 m. However, for studies of the adaptability of the USLE to an area or country with little erosion data, the standard length of 22.1 m (72.6 feet) is best because the equation is normalized on that length. Mutchler (15) provided guidelines for researchers inexperienced in the use of erosion plots.

Anyone beginning experiments to adapt the USLE to a particular situation should start with a set of treatments including continuous fallow and prevailing cropping treatments. Such a set of plots, operated with common procedures, should be established in each major climatic area of a country. These plots will furnish the essential data to adapt and establish confidence in the EI concept (or some other concept). Standard plots and treatments are necessary to correlate the data from different locations and to establish a relationship between rainfall erosivity and soil erodibility. There is no shortcut to such basic data collection, which should be the first endeavor in adapting the USLE to a new geographical area or in developing any erosion equation.

Experiments to acquire data for the L, S, and C factors do not require as much time as the R and K correlations. Such experiments can, therefore, be delayed. Of prime importance is the fact that one set of plots is only a sample of the climate and land conditions at that particular site. Data from a single location must be combined with data from other locations to permit researchers to make erosion predictions over a large area. If such data are not available from other locations, field data must be used to verify a soundly based theoretical prediction method. Of course, a compromise usually is made. There are never enough funds to sample every specific climate-soil combination, and every theoretical equation should be verified across the range of its use to enable interpolation, which is much safer than extrapolation.

Following are more detailed guidelines for evaluating each of the USLE factors:

Rainfall erosivity and soil erodibility. These variables must be evaluated together because they are essentially the cause and effect of soil erosion.

We leave to others the argument of whether the best parameter to represent rainfall erosivity is EI used in the USLE, kinetic energy, momentum, or some other measurement. Some parameter must be used to simplify the highly variable nature of rainfall distribution over even relatively small areas of 25,000 to 50,000 ha. For instance, the R factor (average annual accumulated EI) increases from 5,100 MJ·mm/(ha·h) to more than 9,300 MJ·mm/(ha·h) in the State of Mississippi in the United States over a north-south distance of 480 km. In general, R changes even more where major elevation differences occur. Furthermore, rainfall initiates erosion, so erosion can be predicted only as well as one can predict rainfall and its parameters. Thus, a fairly intensive, well-distributed system of raingages throughout a country is essential for developing a reliable system of assessing the erosion hazard.

For the USLE, the standard erosion surface for measuring the effect of rainfall is a continuously cultivated fallow plot, 22.1 m (72.6 feet) long, that allows development of rill and interrill erosion patterns. This surface was specified in "Instruction for Establishment and Maintenance of Cultivated Fallow Plots," a letter to field workers from D. D. Smith, dated January 1, 1961. The instruction said, "Plow plot to normal depth and *smooth immediately* by disking or cultivating two or more times except for fall plowed plots in areas where wind erosion during the winter may be serious. In such cases delay disking or cultivation until spring. Plowing shall be done each year at the time continuous row crop plots are plowed. Cultivate at row crop planting and cultivating times, in the fall at small grain seeding time, and at other times when necessary to eliminate a serious crust formation. Chemical weed control may also be necessary." The letter also emphasized the requirement that the fallow plot used for computing K be operated according to a "standard plan" by each research location having erosion plots. These plots served to allow national analysis of erosion data from all locations as well as for determining K for the soil on the plot.

Farmers in the United States plow much less than they did in 1961, and the plowing referred to is not used at all in some parts of the world. However, the requirements of a standard treatment and plot to establish K factors and allow convenient comparison of erosion data from different locations remains fully justified.

Measurements of runoff and sediment concentrations from aliquot samples are used to compute soil loss. Samples to determine sediment concentration in the runoff are collected after each storm or each group of closely spaced storms if there is not time to collect samples before the next storm.

Concurrent rainfall measurements are made for each segment of erosion data using a recording raingage. The raingage chart gives rainfall amount per time interval, which is the basis for calculating storm EI.

Slope length (L) and steepness (S). These two factors represent plot topography. For experimental purposes, the plot used for determining the L and S factors, as well as the K factor, should be a plane surface. Such sites may be difficult to find. Also, it is desirable to collect data on slopes of 9 percent for L and slope lengths of 22.1 m (72.6 feet) for S to avoid adjustments in slope data from slope lengths different than 22.1 m long, etc. However, such ideal plot sites are almost never available, especially considering the need for uniform soil conditions.

Slope length experiments must be conducted on plots of several different slopes because L is a function of slope length, λ, and slope steepness, θ. Because slope lengths (as defined for use in the USLE) in nature are relatively short on steep slopes and longer on less steep slopes, some study of the land's topography in an area or country should be made. Terraces and other structures must be used to prevent destruction of soils on slopes exceeding 10 to 12 percent. Therefore, slope length on steep slopes is further restricted. Slope lengths on less steep slopes are much shorter than first appearance would indicate because typical slope irregularities may pond water and invalidate the USLE. Conversely, flatlands often are smoothed or land-planed to a uniform slope. In such cases, slope lengths can exceed 300 m.

Farmland slopes generally are restricted to those that can be farmed with machinery or that will not produce excessive soil loss. There are, of course, notable exceptions to this rule, such as the steep Palouse region in the northwestern United States and mountainside slopes farmed with hand tools.

To repeat, slope lengths and steepnesses of concern must be determined from a study of the location topography and research needs. Plot widths should be the same for all plot lengths or slopes. The difficulty of finding the same soil conditions on different slope steepnesses requires either slope studies within a relatively small range of steepness or an adjustment of soil loss measurements using K for the soil on the plot.

Slope factors can be studied with plots using any cover. However, a fallow surface results in larger amounts of soil loss, which are easier to measure accurately. Use of rainfall simulators is convenient for experiments to determine the L and S factors. As is always the case with rainfall simulation, care must be taken that the simulated storms adequately represent rainstorms received throughout the year at the locations where the research results are to be applied.

Cover and management. The C factor is "the ratio of soil loss from an area with specified cover and management to that from an identical area in tilled continuous fallow" (27). C factors are measured most accurately using the experimental treatment paired with a standard fallow plot. If previous research has established confidently the values of K, L, and S, the soil loss ratio and C factor can be computed by solving the USLE (P is always unity for a standard plot) and the fallow plot is not needed. However, if the research uses simulated rainfall, the fallow plot should be used to reduce the importance of verifying the accuracy of simulated storms.

Cover, canopy, and tillage differ throughout the year. Dividing the year into cropstages allows the use of an average soil loss ratio to replace a relatively small range of values. As a result, the term "C factor" is reserved for the average annual soil loss ratio, and cropstage values are called soil loss ratios. This procedure requires year-around measurements of cover, canopy, and tillage operations. Also, to allow better use of the soil loss ratio at other locations, researchers should measure the cover and canopy of weeds in the treatment. Effects of weeds then will be introduced at the location of the USLE estimate.

Cropstage soil loss ratios and average annual EI for each cropstage are used to compute expected average annual soil loss. This requires research values of the same parameters, except EI values are measured.

Rainfall simulators can be used to evaluate new cropping systems. However, simulators generally cannot be used to evaluate cold weather cropstages. This may be a handicap, depending upon climate and crop. For instance, in Mississippi, cover for row crops is lowest during the winter and early spring, but EI is nearly as high as in the warm season. Also, runoff is highest from the cool, wet soils during the cool season.

Support practices. These variables cannot be evaluated on the standard USLE plot. All rows on standard plots must be parallel to the slope because plot widths are too small to allow the natural runoff concentration and row break-over found in fields.

We have used contour rows on 1,000-m² plots, 22-m slope length, and 46-m row length across the slope. In this case, the topography in the area was such that a 46-m contour length allowed a sufficient distance to approximate the spacing between rills resulting from row break-over. Concrete lining was necessary to prevent deposition in the 46-m-long channel leading to the runoff measuring apparatus. Four of these plots on a 5 percent slope, two on a 2.5 percent slope, and two on a 10 percent slope were instrumented in 1957—all but the 10 percent plots are still in use. These plots were difficult to locate and required some land shaping to achieve the desired plane surface.

In general, it is expensive and difficult to establish experiments using terraces, stripcropping, or other cropping practices. Because of the space required and the variability in topography and soils, such research sites defy efforts to make statistical replication. The support practices, in most cases, are evaluated better on small watersheds.

Analysis. Outlining the proposed analysis of plot experiments can raise questions that cause a revision of procedures in a research outline.

R factor. Several equations are available for describing the erosivity of rainfall. McGregor and Mutchler discussed rainfall energy characteristics (8, 10) and geographical differences in rainfall (19). Wischmeier and Smith (27) truncated the energy-intensity equation formulated by Laws and Parsons (6) to read:

$$KE = 916 + 331 \log_{10} I; \text{ for } I \le 3 \qquad [1]$$

and

$$KE = 1,074; \text{ for } I > 3 \qquad [2]$$

where KE is kinetic energy in foot-tons per acre-inch and I is intensity in inches/hour. Thus, the KE values dependent upon rainfall intensity are close to those based on the equation derived by McGregor and Mutchler (8):

$$KE = 1,035 + 822 \exp(-1.22\ I) - 1,564 \exp(-1.83\ I) \qquad [3]$$

Equations 1, 2, and 3 can be converted to metric units as described by Foster and associates (5).

The major point of the Mutchler and McGregor study (19) was that no known erosion study had distinguished between the storm types that occur throughout the year at any location. The magnitude of storm variations at a location and among locations is sufficient to justify the use of a lumped annual data set, such as that used by Laws and Parsons (6) or McGregor and Mutchler (8).

Rainfall relationships provided by Wischmeier and Smith (27) in *Agriculture Handbook* 537 allow comparison of new research data with data previously collected and used as a basis for the USLE. Our comparison of equations 1 and 3 using 19 years of data from 15 to 34 raingages resulted in almost identical annual average EI values. The restriction of equation 2 had little effect on average EI for our rainfall conditions.

Because of large variations in storm EI, magnitude, and time of year, EI should be measured for every soil loss measurement. For the same reason, at least 3 years of data should be used to evaluate soil loss from an experimental treatment. If the aggregate EI is not representative in magnitude and distribution of the values in *Agriculture Handbook* 537, the project should continue for another year. At locations without established EI distributions, an analysis of measured rainfall distributions can be made with somewhat lesser confidence to determine the representativeness of the rainfall during the experiment. This same problem is magnified when using rainfall simulation to evaluate soil erodibility because soil loss is affected by time of year (temperature, antecedent moisture, etc.) in addition to storm EI.

K factor. Soil erodibility is "the soil loss rate per erosion index unit for a specified soil as measured on a unit plot..." (27). Measurements on a unit plot mean measurements made over a sufficiently long period to sample adequately the rainfall at the location. Wischmeier and Smith (27) suggested using 22-year records to compute R because of the cyclical patterns in rainfall. This restriction, however, is only a guideline. As mentioned earlier, 3 years of rainfall data usually represent the annual rainfall distribution adequately in northern Mississippi, where rainfall roughly averages 1,400 mm/y, ranges from 1,000 to 1,800 mm/y, and is distributed fairly evenly throughout the year except during the somewhat drier late summer and fall months. In addition, the rain falls on soil that is seldom frozen during rainfall. The analysis of rainfall (and EI) representation for areas that have lower rainfall amounts and large portions of the year with frozen soil requires more years of replication. Low annual rainfall would be treated more easily and more properly on a storm basis.

The K factor is a measure of the effect of rainfall and runoff on soil erosion, as measured throughout the year. A particular soil may rill more easily than another. Rain may arrive at different times after the prescribed tillage on the standard USLE plot. It may arrive at different antecedent water contents of the soil. It does arrive over a year of varying temperature. All of these factors affect the amount of erosion resulting from a unit of EI. Thus, K is not entirely a measure of the inherent erodibility of soil. K is properly computed as follows:

$$SL = a + K (EI) \quad\quad\quad [4]$$

where SL is storm soil loss, EI is the storm parameter of energy computed with equations 1 and 2 and maximum 30-minute intensity, and a is a constant that represents primarily the effect of soil water content and temperature at the beginning of rainfall. The restrictions on E and I given in *Agriculture Handbook 537* (27) and outlined below attempt to reduce a to zero to attain the proportionality of A, R, and K.

A storm is defined arbitrarily as one rain separated from another rain by more than 6 hours with less than 1 mm (0.04 inch) of rain. Storms with less than 13 mm (0.5 inch) of rain generally do not cause appreciable soil loss, defined as greater than or equal to 0.02 t/ha (0.01 ton/acre). However, short, intense bursts of rainfall less than 13 mm often exceed the infiltration rate of a plot soil and cause measurable soil loss. Therefore, rains of less than 13 mm but with a 15-minute intensity of 25 mm/h (1 inch/hour) or greater are included in the EI/SL relationship. Also, I30 in EI is limited to 64 mm/h (2.5 inches/hour). However, we always use small EI data if there is any measurable soil loss from the storm.

These criteria are suitable for general application of the USLE. However, they should be verified at any location or situation outside the data base of the USLE.

In our experimentation, we consider the above criteria as factual and use the following equation:

$$K = b \ \Sigma SL / \Sigma EI \qquad\qquad [5]$$

where b represents the other factors of the USLE that are not unity.

Because K-factor evaluation is a time-consuming process, few soils have been evaluated using rainfall. Instead, rainfall-evaluated soils have been used as a basis for evaluations using rainfall simulators. A fallow plot is relatively easy to prepare, but a careful correlation of simulated rainfall data with rainfall must be made to obtain confidence in such measured K factors. Wischmeier and associates (25) used a weighting procedure for accumulating the results of rainfall simulator storms of 0.5, 1, 1.5, and 2 hours to account for storm patterns of natural rainfall and antecedent soil water conditions in the U.S. Corn Belt. Other researchers have used different methods, but any K-factor evaluation with simulated rainfall must consider rainfall distribution at a location.

Using simulated rainfall is complicated by the fact that the K factor is not a constant but varies throughout the year (16). The K factor of a Mississippi soil varies from a low of 31 percent of annual average on August 5 to a high of 169 percent on February 4. For this reason, it is desirable to have K-factor plot data to correlate simulated rainfall results.

L factor. The slope length factor "is the ratio of soil loss from the field slope length to that from a 22.1-m (72.6-foot) length under identical conditions" (27). To the researcher, this definition of L also means that the plot widths of slope-length plots should be the same.

Past research has shown the L-factor relationship to be

$$L = (\lambda/22.1)^m \tag{6}$$

where λ is slope length in meters and m is a function of slope steepness, θ, in degrees. Thus, m is the only variable to be evaluated in a slope-steepness experiment, and formulation of

$$m = f(\theta) \tag{7}$$

requires slope-length plots on several different slopes. Data from a set of slope-length plots can be used to fit

$$A(\lambda) = a\lambda^m \tag{8}$$

where $A(\lambda)$ is the soil loss from each plot and a is a proportionality constant depending upon all of the USLE factors except slope length. By definition,

$$L = A(\lambda)/A(22.1) \tag{9}$$

However, results from a 22.1-m-long plot are not required because m can be evaluated from equation 8 with other plots of various lengths.

The exponent m, as given in *Agriculture Handbook 537* for use by field technicians, is a set of discrete values for a range of slopes. Mutchler and Murphree (*17*) used data from Mutchler and Greer (*18*) and from earlier research to derive the following:

$$m = 1.2(\sin\theta)^{1/3} \tag{10}$$

which produced a more accurate exponent for slopes under 2 percent and made little change in the exponents given in *Agriculture Handbook 537* for steeper slopes.

Although crops can be grown on slope-length plots, a fallow surface tilled similarly to the continuous fallow treatment is best. The higher rates of erosion under fallow conditions are advantageous because they are easier to measure accurately. This same observation applies to slope-steepness plots.

Foster and Wischmeier developed a procedure (*4, 27*) to evaluate slope lengths that are slightly irregular.

S factor. The slope-steepness factor "is the ratio of soil loss from the field slope gradient to that from a 9 percent slope under otherwise identical conditions" (27). Evaluation or verification of the S factor is done most easily on standard plots with a fallow surface. Erodibility will be near the K value of the soil, C will be near unity, and L will be unity. Thus, data from the slope plots can be used to fit the parameters of

$$A(\theta) = f(\theta) \tag{11}$$

and

$$S = A(\theta)/A(9\%) \tag{12}$$

The first slope relationships were given in terms of percent slope because it was and is a convenient field measurement. The relationship given in *Agriculture Handbook 537* is based upon slope in degrees:

$$S = 65.41 \sin^2\theta + 4.56 \sin\theta + 0.065 \tag{13}$$

This equation evolved from Smith and Wischmeier (20):

$$A = 0.43 + 0.30 S + 0.043 S^2 \tag{14}$$

where this S is percent slope. Equation 14 was normalized by dividing by the value of soil loss, A, for S = 9 percent. Later, researchers decided to substitute $\sin\theta$ for S because percent slope is 100 tan θ and tan θ gave unrealistic values at high angles. This change was primarily theoretical because the sine and tangent are not much different for the slope where the USLE has validity.

The above slope equation was developed using data generally from 3 to 18 percent slopes. Murphree and Mutchler (14) found that equation 13 overestimated soil loss from slopes less than 3 percent. They proposed a slope factor coefficient of

$$S_c = (\beta/2) + 0.5 \tag{15}$$

where

$$\beta = 1 - 0.67 \exp(-160 \sin\theta) \tag{16}$$

to modify the S factor at low slopes. The experimental work was done by grading a 0.2 percent slope to slopes of 0.1, 0.2, 0.5, 1, 2, and 3 percent while preserving the A horizon (about 8 inches deep) on all plots. They used the rainulator (11) to produce runoff and erosion from the plots. It is, of course, preferable to use plots without any grading; smoothing is usually a necessity, especially on less steep slopes.

McCool and associates (7) developed the following equation:

$$LS = \left(\frac{\lambda}{22.13}\right)^{0.3} \left(\frac{s}{9}\right)^{1.3} \qquad [17]$$

where s is percent slope for slopes of 9 to 60 percent in the dryland grain region of the northwestern United States. They used data from field observations and erosion plot data to formulate the equation.

Experiments to collect data for a generalized S factor are difficult, primarily because the same soil cannot be found on different slopes.

C factor. The cover and management factor "is the ratio of soil loss from an area with specified cover and management to that from an identical area in tilled continuous fallow" (27). Evaluations of cropping systems can be made using erosion plots under rainfall or simulated rainfall. In both cases, the best method is to compare a cropping treatment and a fallow treatment on identical plots. Thus,

$$SLR = A(\text{treatment})/A(\text{fallow}) \qquad [18]$$

The C factor for a particular cropping system is computed from soil loss ratios (SLR) that are computed using equation 18 for a cropstage that is a period of time when "cover and management effects may be considered approximately uniform" (27). Cropstages are defined in *Agriculture Handbook 537*. The researcher measures SLR for the numbers (n) of cropstage periods, and the USLE application uses the following:

$$C = \sum_{i=1}^{n} (EI_i \times SLR_i) / \sum_{i=1}^{n} EI_i \qquad [19]$$

Equation 19 implies that SLRs are constant for all parts of a country or area of USLE usage and accounts for different EI distributions at different locations within the country.

Standard-size erosion plots under rainfall are used in blocks with one fallow plot and several cropping treatments. In Mississippi, 3 years or more of replicated data are required, depending upon an evaluation of the rainfall received, to evaluate a cropping system properly. Areas with less rainfall may require more annual replications to sample the expected EI or rainfall distribution properly.

Recently, there has been increased interest in further understanding the C factor using subfactors based on suggestions by Wischmeier (24). Consequently, we routinely estimate cover and canopy throughout the year, record observations of surface configuration and tillage application, and record cropping history. Also, we measure crop yield and details of residue management to describe SLRs for the cropping system more properly. Because weed varieties and populations differ by location, we measure cover and canopy separately (9) to allow reporting of SLRs without weeds. The ratios can then be adjusted for expected weeds by the USLE user.

Use of a rainfall simulator to determine SLRs requires careful analysis to represent properly the EI and crop growth of a particular cropstage.

Outdoor use of rain simulation is limited to the warm seasons of the year; simulators do not operate well in cool, wet conditions and not at all in freezing temperatures. We have used a rainulator primarily for slope studies, C-subfactor experiments, and related experiments on the mechanics of erosion.

Use of a fallow plot to evaluate SLRs with equation 18 requires that the fallow plot be prepared at least 2 years prior to the comparison with treatment plots to eliminate the effects of previous cropping. Also, fallow plots must not be used too long because the soil will deteriorate from excessive erosion and, hence, will not properly represent the soil used for adjacent cropping treatments. There are two solutions to this problem: either use the fallow plot only long enough to determine K for the soil or use the SLRs for a carefully evaluated crop that will not significantly change the soil erodibility.

The SLR can be evaluated using a known K with the following equation:

$$ SLR = \frac{\Sigma SL}{K \; L \; S \; P \; \Sigma EI} \qquad [20] $$

If a standard plot is used, L and P are unity (S usually is not unity). The SLR is computed by dividing the accumulated soil loss by the accumulated EI for the particular cropstage (replicated by years) and by a constant representing K and S.

If a check plot (with a known C) is used, then the equation is as follows:

$$ SLR = \frac{\Sigma \; SL(\text{treatment})}{\Sigma \; SL(\text{check})} \times \frac{SLR(\text{check})}{1} \qquad [21] $$

In our research, we use a check plot to increase confidence in our results even if we know the soil erodibility.

P factor. The support-practice factor "is the ratio of soil loss with a support practice like contouring, stripcropping, or terracing to that with straight-row farming up-and-down the slope" (27). As discussed earlier, support practices cannot be evaluated using standard USLE plots.

The effectiveness of support practices are difficult to evaluate. Consequently, there are few data available for this purpose. P factors must be evaluated on slope widths that will evaluate row break-over. The proposed analysis of our slope-segment plots (Figure 3) depends upon data from standard plots with the same crop on both types of plot. Thus,

$$P \text{ (contouring)} = \frac{A(\text{contour})}{A(\text{plot})} \qquad [22]$$

which follows the definition given earlier because both plots have the same slope length. However, adjustments must be made for any differences in slope and soil.

Tillage treatments, ranging from relatively smooth surfaces to row

Figure 3. A contour segment plot (46-m contour rows and 22.1-m slope length) used to determine the reduction in soil loss due to tilling and planting on the contour. The plot is bordered on the sides and top by earthen ridges. A concrete-lined channel along the bottom of the plot carries runoff to the runoff collector and H-flume equipped with a water-stage recorder. Sediment concentrations in the aliquot sample taken with the Coshocton sampler are used with measured runoff to compute storm soil loss.

ridges, impose various surface configurations on a field. The combination of these configurations with slope length, steepness, and topography makes it almost impossible to design a research program to evaluate contouring comprehensively. The same is true of strip-cropping. Consequently, present-day P-factor values for these practices have been systematized and evaluated primarily by group agreement among researchers and USLE users and with very little data.

A contour plot, like our slope segment, should carry runoff out of the end of the rows (side of the plot) for smaller rains and through the interior of the plot at some low point in the rows for a storm sufficiently large to cause row break-over. After row break-over, most runoff would exit at that point until subsequent tillage rebuilt the row ridges. Solving the row slope design problem would, of course, provide the answer or bias the answer to the contour problem. The best solution to this dilemma is to use well-described watersheds and adequate annual replications to compare contour and up-and-down-hill tillage.

Large plots—unit-source watersheds

The third type of erosion research plot is a small catchment or watershed large enough to include at least one natural drainageway. These often are called "unit-source watersheds," which also implies that only a single crop is grown on the entire catchment. This plot would contain interrill areas, rills, and small ephemeral channels that may or may not be cultivated or planted to the same crop as the remainder of the catchment area. Also, terraces or other practices may be included in this large plot. The attractive feature of this plot is that it may combine the results of all the erosion processes and conservation measures in a single measurement. However, use of large plots affords little opportunity to learn about the different parts of the erosion process. Also, natural catchments are not replicated easily to increase confidence in the results.

Some support practices affecting erosion require so much space that they are difficult to evaluate on any plot smaller than a small unit-source watershed of 2 to 4 ha. Contouring and stripcropping are among such practices. Terraces can be evaluated on watersheds, but because they concentrate and control runoff, they are evaluated better as a simple terrace interval unit. Small watersheds have a major attribute—farmers are more willing to believe results from larger scale experimentation. They also have a major disadvantage—research results on one watershed are difficult to apply to other watersheds.

Small watersheds are best for research that seeks to evaluate conservation systems or to verify a modeling concept. Such research is justified by the need for basic data to give the researcher confidence that formulations are correct and to persuade conservation workers that conservation systems effectively conserve soil and are usable by farmers.

Procedures. Site selection is critical for research with small watersheds because they must be representative of a significant area of land. However, the watershed characteristics must still be measured carefully. In contrast to plots used to study isolated parts of the erosion process, small watersheds allow study of the entire process as affected by topography and climate. Transfer of the results to unmeasured watersheds and verification of models then can be made more easily from a comprehensive description of the research watershed.

As with field plots, a recording raingage record of rainfall is needed to compute rainfall intensities and energy and for analysis of rainfall distribution. Limits of a watershed should be established with some type of border. We usually use a low earthen ridge turned up with a farm implement.

Runoff and sediment yield from small watersheds are too great for continuous sampling. On flatland watersheds, we have used a Parshall flume and a water-stage recorder to measure runoff. A pumping sampler can extract sediment concentration samples at preset time intervals, usually 5 minutes initially, until storm analysis indicates that a larger interval is more appropriate. On small upland watersheds, we have used an H-flume and a water-stage recorder with an intermittent proportional sampler to obtain sediment concentrations. Details of instrumentation can be found in *Agriculture Handbook 224 (2)*.

The size of the flume should be determined on the basis of the largest runoff rate expected because larger storms produce the largest amounts of sediment. For USLE-type plots, the design runoff rate used is 100 percent of the 5-minute-duration rainfall intensity. For larger plots or unit-source watersheds, rainfall duration can be estimated from the estimated time of concentration for the watershed.

Some local experience in estimating runoff rates is highly desirable.

All details of cropping and management practices should be recorded, including estimates of residue cover, weed cover, and crop canopy, at sufficient times during the year to allow a description of changes. The USLE factors can be used as a guideline for describing the characteristics of the farmable part of a watershed. The efficiency of sediment transport from interrill areas and the erosion of rills and small channels becomes more important for watersheds. Thus, researchers should pay particular attention to drainage and rill pattern characteristics.

Sediment analysis of the concentration samples is done by routine laboratory methods. Both total sediment and particle-size distribution analyses should be made. Particle size is important in sediment transport and deposition.

Most data collection problems arise because the equipment is not ready to operate when runoff occurs. Therefore, gaging stations should be inspected at least once a week. Operability of the samplers should be verified after a storm. However, seemingly unexplainable problems can arise between storms. All of our watershed equipment is automatic, but there is no substitute for the attention of a good technician to keep the equipment operable.

Analysis. Data from small watersheds are analyzed on a storm basis. We use the same storm definition for watershed plots as for USLE plots. To determine storm sediment yield, runoff increments between concentration samples are multiplied by the average sediment concentration to obtain sediment yield increments, which are then accumulated to determine storm sediment yield. Missing data are a particularly difficult problem because watersheds are so difficult to replicate. If a runoff measurement is lost, the storm sediment yield is estimated from examination of the long-term record. If all the sediment samples are lost, the runoff can be used to help estimate the storm sediment yield. More commonly, part of the sediment concentration samples may be lost due to sampler malfunction. If most of the sediment samples are available, they can be used to construct a smooth sediment concentration graph to estimate the missing values by interpolation. Of course, if only representative storm results are desired, the storm with too many missing measurements is discarded.

Small watersheds usually have a high SDR. Little eroded material is deposited in depositional areas. Therefore, small watersheds may be used to evaluate factors of the USLE with little error. The SDR can be estimated by inspecting the watershed for evidence of deposition. SDR also can be estimated by analyzing the primary particle size of the watershed soil and of the sediment yield. By assuming that sediment is eroded uniformly from the watershed surface and that all clay leaves the watershed in the runoff, the amount of silt and sand deposited within the watershed can be computed. This approach is feasible only where the SDR is about 90 percent or greater.

Researchers can use several procedures on small watersheds to evaluate conservation practices. A single watershed can be operated with a calibration treatment for several years, then subsequently with some treatment, such as contouring. The change in sediment yield, adjusted for rainfall differences, then is the treatment effect. Also, two watersheds with different treatments can be paired and operated several more years. Both of these procedures are time-consuming, and some changes in watershed performance due to previous erosion must be accounted for. A better system is to use a single watershed to verify a model. Use of a model is quicker and often enhances use of the experimental results on unmeasured areas. Inherent in the modeling approach is the requirement of some experience to use as the basis for forming a concept of the watershed sediment yield process.

Conclusions

Erosion research includes studies using plots representative of the landscape or part of the landscape that is of interest. A carefully written research outline ensures that the study focuses on a well-defined part of the erosion process and that an analysis of the data can provide either basic knowledge useful for further research or for application to erosion problems in the field.

Three types of erosion plots are useful—small plots used primarily for studying interrill erosion; USLE standard plots that include both rill and interrill processes; and the small, unit-source watershed that includes rill, interrill, deposition, and small channel processes.

Small plots probably are most useful for establishing principles and accumulating knowledge about rainfall-induced erosion for further use in appplied research. Experiments using small plots can be controlled more tightly than experiments with large plots. This allows studies of isolated variables. Some problems with research using small plots are created by the large spatial variability of soil and rainfall.

The USLE standard plot is perhaps the most important plot because of the extensive U.S. data base collected generally in a uniform manner and the proven usefulness of the USLE to soil conservationists. Also, USLE plots are convenient for use as a reference in discussing the principles of field-plot experimentation.

Small, unit-source watersheds are the only type of plot large enough to evaluate conservation structures and practices, such as contour rows. Being difficult to replicate, they are most useful for verification of runoff and sediment yield models.

These three types of plots are by no means all those used for the many diverse objectives of research in soil erosion.

REFERENCES

1. Al-Durrah, M., and J. M. Bradford. 1981. *New methods of studying soil detachment due to waterdrop impact.* Soil Science Society of America Journal 45: 949-953.
2. Brakensiek, D. L., H. B. Osborn, and W. J. Rawls. 1979. *Field manual for research in agricultural hydrology.* Agriculture Handbook 224. U.S. Department of Agriculture, Washington, D.C. 550 pp.
3. Dissmeyer, G. E., and G. R. Foster. 1980. *A guide for predicting sheet and rill erosion on forest land.* Technical Publication SA-TP 11. Forest Service, U.S. Department of Agriculture, Washington, D.C. 40 pp.
4. Foster, G. R., and W. H. Wischmeier. 1974. *Evaluating irregular slopes for soil loss prediction.* Transactions, American Society of Agricultural Engineers 17: 305-309.
5. Foster, C. R., D. K. McCool, K. G. Renard, and W. C. Moldenhauer. 1981. *Conversion of the universal soil loss equation to SI units.* Journal of Soil and Water Conservation 36(6): 355-359.
6. Laws, J. O., and D. A. Parsons. 1943. *The relation of raindrop size to intensity.* Transactions, American Geophysical Union 24: 452-460.
7. McCool, D. K., W. H. Wischmeier, and L. C. Johnson. 1982. *Adapting the universal soil loss equation to the Pacific Northwest.* Transactions, American Society of Agricultural Engineers 25: 928-934.
8. McGregor, K. C., and C. K. Mutchler. 1976. *Status of R-factor in North Mississippi.* In *Soil Erosion: Prediction and Control.* Soil Conservation Society of America, Ankeny, Iowa. pp. 135-142.
9. McGregor, K. C., and C. K. Mutchler. 1983. *C factors for no-till and reduced-till corn.* Transactions, American Society of Agricultural Engineers 26: 785-788, 794.
10. McGregor, K. C., C. K. Mutchler, and A. J. Bowie. 1980. *Annual R values in North Mississippi.* Journal of Soil and Water Conservation 35: 81-84.
11. Meyer, L. D., and D. L. McCune. 1958. *Rainfall simulator for runoff plots.* Agricultural Engineering 39: 644-648.
12. Meyer, L. D., and W. C. Harmon. 1979. *Multiple-intensity rainfall simulator*

for erosion research on row sideslopes. Transactions, American Society of Agricultural Engineers 22: 100-103.

13. Moldenhauer, W. C. 1965. *Procedure for studying soil characteristics using disturbed samples and simulated rainfall.* Transactions, American Society of Agricultural Engineers 8(1): 74-75.
14. Murphree, C. E., and C. K. Mutchler. 1981. *Verification of the slope factor in the universal soil loss equation for low slopes.* Journal of Soil and Water Conservation 36: 300-302.
15. Mutchler, C. K. 1963. *Runoff plot design and installation for soil erosion studies.* ARS-41-79. Agricultural Research Service, U.S. Department of Agriculture, Washington, D.C.
16. Mutchler, C. K., and C. E. Carter. 1983. *Soil erodibility variation during the year.* Transactions, American Society of Agricultural Engineering 26: 1,102-1,104, 1,108.
17. Mutchler, C. K., and C. E. Murphree. 1981. *Prediction of erosion on flatlands.* In R.P.C. Morgan [editor] *Soil Conservation, Problems and Prospects.* John Wiley & Sons, Chichester, England. pp. 321-325.
18. Mutchler, C. K., and J. D. Greer. 1980. *Effect of slope length on erosion from low slopes.* Transactions, American Society of Agricultural Engineers 23: 866-869, 876.
19. Mutchler, C. K., and K. C. McGregor. 1979. *Geographical differences in rainfall.* In *Proceedings, Rainfall Simulator Workshop, Tucson, Arizona, March 7-9, 1979.* ARM-W-10. U.S. Department of Agriculture, Washington, D.C. pp. 8-16.
20. Smith, D. D., and W. H. Wischmeier. 1957. *Factors affecting sheet and rill erosion.* Transactions, American Geophysical Union 38: 889-896.
21. U.S. Department of Agriculture. 1979. *Field manual for research in agricultural hydrology.* Agriculture Handbook No. 224. Washington, D.C. 550 pp.
22. U.S. Department of Agriculture. 1980. *CREAMS—A field scale model for chemicals, runoff, and erosion from agricultural management systems. Volume I: Model documentation and Volume II: User manual.* Conservation Research Report No. 26. Washington, D.C.
23. Williams, J. R. 1975. *Sediment-yield prediction with universal equation using runoff energy factor.* In *Present and Perspective Technology for Predicting Sediment Yield and Sources: Proceedings, Sediment Yield Workshop, USDA Sedimentation Laboratory, Oxford, Mississippi, November 28-30, 1972.* ARS-S-40. U.S. Department of Agriculture, Washington, D.C.
24. Wischmeier, W. H. 1975. *Estimating the soil loss equation's cover and management factor for undisturbed area.* In *Present and Perspective Technology for Predicting Sediment Yield and Sources: Proceedings, Sediment Yield Workshop, USDA Sedimentation Laboratory, Oxford, Mississippi, November 28-30, 1972.* ARS-S-40. U.S. Department of Agriculture, Washington, D.C.
25. Wischmeier, W. H., C. B. Johnson, and B. V. Cross. 1971. *A soil erodibility nomograph for farmland and construction sites.* Journal of Soil and Water Conservation 26: 189-193.
26. Wischmeier, W. H., and D. D. Smith. 1965. *Predicting rainfall erosion losses from cropland east of the Rocky Mountains.* Agriculture Handbook 282. U.S. Department of Agriculture, Washington, D.C.
27. Wischmeier, W. H., and D. D. Smith. 1978. *Predicting rainfall erosion losses.* Agriculture Handbook 537. U.S. Department of Agriculture, Washington, D.C. 58 pp.

3

D. E. Walling

Measuring sediment yield from river basins

Information on the sediment yield at the outlet of a river basin can provide a useful perspective on the rates of erosion and soil loss in the watershed upstream. Many countries undertake such measurements of sediment transport as part of national hydrometric programs or more specific research investigations. Therefore, some background data are often available. In nearly all cases, these measurements refer to the suspended sediment load of the river. The bedload component is not included because of the practical difficulties in obtaining such measurements (26). Where data are unavailable, measurement programs can be initiated using standard equipment and procedures (40).

A global perspective

On a global basis, documented suspended sediment yields include minimum values well below 2 t/km²/yr. For example, Douglas (21) cites yields of 1.3 t/km²/yr for the Brindabella catchment (26.1 km²) and 1.7 t/km²/yr for the Queanbeyan River (172 km²) in the Southern Tablelands and Highlands of New South Wales, Australia. Values of less than 1.0 t/km²/yr have been reported for several rivers in Poland (11).

Improved availability and global coverage of sediment yield data in recent years have significantly changed the perception of the upper bounds of sediment yield. Values in excess of 10,000 t/km²/yr have been reported for several rivers (Table 1). The highest value shown in table 1 is for a small, unnamed river in northeast Taiwan included on a map produced by Li (49). However, some uncertainty surrounds this value. A leading contender for the global maximum value must be the Dali River in North Shaanxi in the People's Republic of China, with a yield of 25,600 t/km²/yr from a basin area of 96.1 km² and 16,300 t/km²/yr from a 3,893-km² basin (60). The Dali, a tributary of the middle reaches of the Yellow River, drains the gullied loess region, an area known for its severe erosion problems. Converted to local rates of soil loss, the sediment yields of the Dali River represent soil erosion rates on the order of 250 t/ha/yr.

Sediment yield data are available for many areas of the world, but the global coverage is inadequate to produce a reliable and detailed world map of sediment yields. Figure 1 nevertheless represents a tentative map I prepared of the global pattern of sediment yield from river basins on the order of 1,000 km² to 10,000 km² in area. The pattern, discussed in detail elsewhere (99), reflects the interaction of climatic, topographic, geologic, and other physiographic controls and provides some indication of the global distribution of fluvial erosion.

There are, however, several major problems with any attempt to use sediment yield data to provide meaningful information about

Table 1. Recently reported values of sediment yield in excess of 10,000 t/km²/yr.

Location	River	Drainage Area (km²)	Mean Annual Sediment Yield (t/km²/yr)	Source
Shaanxi, China	Dali	96.1	25,600	(60)
	Dali	187	21,700	
	Dali	3,893	16,300	
Kenya	Perkerra	1,310	19,520	*
Taiwan	(unknown)	(unknown)	31,700	(49)
Java	Cilutung	600	12,000	(42)
	Cikeruh	250	11,200	
New Guinea	Aure	4,360	11,126	(71)
North Island	Waiapu	1,378	19,970	
New Zealand	Waingaromia	175	17,340	(37)
	Hikuwai	307	13,890	
South Island	Hokitika	352	17,070	
New Zealand	Cleddau	155	13,300	(36)
	Haast	1,020	12,736	

*Unpublished report from T. Dunne, Department of Geological Sciences, University of Washington, Seattle.

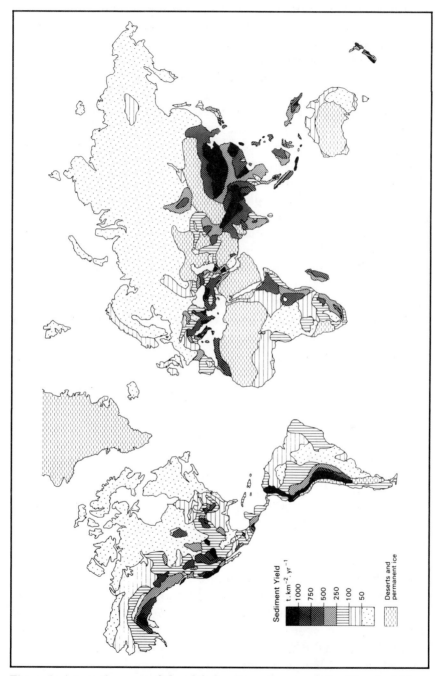

Figure 1. A tentative map of the global pattern of suspended sediment yield.

on-site rates of erosion and soil loss within a drainage basin. The first relates to the processes of sediment delivery or conveyance interposed between on-site erosion and downstream sediment yields (78, 95). Only a fraction and perhaps a rather small fraction of the soil eroded within a drainage basin reaches the basin outlet and is represented in the sediment yield. Deposition and temporary and permanent storage may occur on the slopes, particularly where gradients decline downslope; at the base of a slope; in swales; on the floodplain; and in the channel itself. The relative magnitude of this loss tends to increase with increasing basin size. Hadley and Shown (41), for example, indicated that only 30 percent of the sediment eroded in many of the small tributary basins (0.5-5.2 km²) of the Ryan Gulch Basin in northwestern Colorado reached the main valley. In turn, only 30 percent of this sediment is transported to the mouth of the 124.8-km² basin. Similarly, a countrywide study of 105 agricultural production areas in the United States by Wade and Heady (92) documented a range of sediment output between 0.1 percent and 37.8 percent of gross erosion.

The concept of a sediment delivery ratio (SDR) has been introduced as a means of quantifying these effects (33, 53, 79). SDR is the ratio of sediment delivered at a basin outlet (t/km²/yr) to the gross erosion within the basin (t/km²/yr). SDR values as low as 0.05, or 5 percent, frequently are cited. However, considerable uncertainty surrounds the methods available for calculating or estimating SDR for a drainage basin.

A second problem in attempting to link downstream sediment yield to upstream erosion rates is the temporal discontinuity that may be involved in sediment delivery. Sediment eroded at one location may be temporally stored and subsequently remobilized several times before reaching the drainage basin outlet. Sediment transported out of a basin may therefore reflect the recent history of erosion and sediment delivery rather than contemporary erosion. This situation is clearly exemplified by the work of Trimble (85, 86, 87) in Georgia and in the Driftless Area of Wisconsin in the United States. Severe upland soil erosion occurred within these areas during the late nineteenth and early twentieth centuries, but most of this sediment was deposited in the river valleys. Only about 5 percent of the eroded material reached the basin outlets. Soil conservation measures implemented in the 1930s and thereafter reduced upland erosion about 25 percent but failed to reduce sediment yield significantly because

Table 2. Comparisons of measured suspended sediment yields for African rivers and estimated rates of contemporary soil loss by water erosion depicted on the FAO map of soil degradation (32).

River	Country	Basin Area (km²)	Suspended Sediment Yield (t/km²/yr)	FAO Estimate of Soil Loss (t/km²/yr)
Watari	Nigeria	1,450	483a*	1,000-5,000
Bunsuru	Nigeria	5,900	438a	1,000-5,000
Senegal	Mali	157,400	14.6b	1,000-5,000
Faleme	Mali	15,000	40b	1,000-5,000
Hammam	Algeria	485	198c	1,000-5,000
Kebir Ouest	Algeria	1,130	92c	1,000-5,000
Mesanu	Ethiopia	150	1,680d	5,000-20,000

*Reported by (a) Oyebande (70), (b) Comite InterAfrican détudes hydrauliques, (c) Demmak (16), (d) Virgo and Munro (91).

sediment stored in the valley deposits was remobilized. Meade described a similar situation for other rivers in the eastern United States (57).

A third problem in relating sediment yield downstream to soil erosion upstream derives from the fact that the sediment transported by a river represents material derived from a variety of sources other than upland soil erosion. These sources include channel and gulley erosion and mass movements reaching the channel network. For example, in a study of the Ader Dutchi Massif region in Niger, Heusch (44), using plot studies, estimated local soil erosion rates at about 8 t/ha/yr within a 117-km² watershed. On the other hand, sediment yield from the basin, estimated from reservoir surveys, was about 40 t/ha. Heusch concluded that about 75 percent of the sediment reaching the basin outlet was derived from channel erosion and gullying. If the SDR associated with soil erosion from the slopes of the watershed was less than 100 percent, this value of 75 percent may itself be an underestimate of the contribution derived from the additional sources.

Data in table 2 effectively exemplify the magnitude of the problems involved in relating sediment yield data from river basins to rates of soil erosion in the watershed upstream. Table 2 compares measured sediment yields from a number of African river basins with the estimates of contemporary soil erosion rates within these basins; this information is depicted on a United Nation's Food and Agriculture Organization map of current rates of soil degradation by water erosion in Africa north of the equator (31). In all cases, the estimates of soil erosion rates are about an order of magnitude greater than the reported sediment yields. This discrepancy could be explained in terms of SDR being of the order of 10 percent in these drainage basins. This would be consistent with current views on the likely magnitude of such ratios in sizable river basins. It also suggests that the sediment delivery problem probably represents the major difficulty in any attempt to relate downstream sediment yields to local erosion rates in the watershed upstream.

Reliability of existing sediment yield data

In some locations, appropriate sediment yield data may be available from existing monitoring programs. However, before researchers or others use such data for further research or planning, careful consideration must be given to the data's reliability. A useful, although perhaps extreme, example of the potential magnitude of this problem is provided by recent work on the suspended sediment loads of New Zealand rivers by Griffiths (35) and Adams (2). Both workers used the same basic discharge and concentration data collected by the New Zealand Ministry of Works to estimate the mean annual suspended sediment load of the Cleddau River, which drains a 155-km² basin in the southwestern part of South Island. Both used rating curve procedures to calculate the loads, but their published loads of 13,300 t/km²/yr and 275 t/km²/yr differ by nearly two orders of magnitude.

In some instances, several different estimates of sediment yield may be available from existing sources as a result of the use of different load calculation techniques or different periods of record. For example, several researchers have attempted to estimate the suspended sediment yield of the Upper Tana River in Kenya over a period of 20 years (Figure 2). These estimates range over an order of magnitude. Estimates A-E were derived using essentially the same basic flow and sediment concentration data. Estimates F and G were based on a new sampling program. Comparison of their load estimates for the main river with those for the upstream tributaries prompted Dunne and Ongweny (23) and Ongweny (68) to suggest that the former were underestimates. They produced revised estimates (H and I) by summing and extrapolating the estimates for the individual upstream tributaries. More recent surveys of sedimenta-

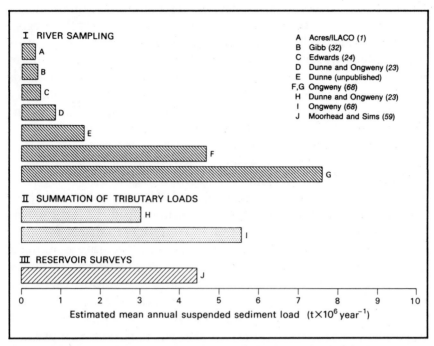

Figure 2. Estimates of the mean annual suspended sediment load of the Upper Tana River, at Kamburu, Kenya, produced by different workers.

tion rates in Kamburu Reservoir by Moorhead and Sims (59), coupled with an estimated trap efficiency for the reservoir of 75 percent, provide an alternative basis for estimating the sediment load. Current opinion would, in fact, indicate that the actual mean annual suspended sediment load of this river probably now exceeds 5 million t/yr, although it is likely to have increased markedly over the past 20 years as a result of land use change.

Walling and Webb (98), Dickinson (18), and Allen and Petersen (4) have discussed factors involved in assessing the reliability of sediment yield data. These include the representativeness of the suspended sediment samples collected from the river channel and the efficiency of laboratory techniques used to determine the concentration of individual samples. Many measurement programs take samples at or near the water surface. Such samples probably underestimate the sediment load of a river, perhaps by as much as 25 percent. Conversely, from a study on Oklahoma's Washita River in the

United States, Allen and Petersen (4) suggested that equal transit-rate sampling (collection of a composite sample by depth integration at several equally spaced verticals across a stream) by operators with limited experience could lead to overestimation of the long-term load by 30 percent. Although national and international standards relating to sampling equipment and techniques as well as laboratory procedures have now been established (39, 45, 105), sediment sampling still involves many problems and uncertainties; careful attention must be given to the likely reliability of the methods used in the monitoring programs associated with existing data.

Greater problems surround the methods used to assess long-term sediment loads and annual sediment yields from intermittent measurements of instantaneous suspended sediment load. These problems center around the marked and rapid fluctuations in suspended sediment concentration exhibited by many rivers. Accurate assessment of suspended sediment loads for specific periods necessitates detailed records of sediment concentration that may be combined with the records of water discharge. Continuous records of stream discharge are available at most measuring stations. But an equivalent record of sediment concentration may be difficult to obtain by a progam of manual sampling. On large rivers, it may be possible to collect sufficient manual samples to define meaningfully the record of sediment concentration during periods of fluctuating concentration, and procedures are available to establish the trend between individual samples (73). On smaller streams, concentrations fluctuate more rapidly during flood events, and operational constraints may limit the frequency at which manual samples can be collected. Attempts have been made to resolve these difficulties by developing equipment and instruments capable of automatic collection of data on sediment concentration. Many researchers have used automatic pump-sampling equipment (90, 97). Continuous-recording turbidity meters (10, 29, 38) and nuclear probes (52, 74, 84) also have been used successfully in a few cases. Difficulties may arise in relating the point values of concentration obtained with such equipment to the mean value for the cross section, but this limitation is frequently of little significance when compared with the positive improvements in temporal resolution obtained.

More often, however, frequent manual samples or additional information from automatic samplers or recording equipment are not available, and the detailed record of sediment concentration cannot be defined. In this case, researchers must use indirect load calculation procedures, which involve either interpolation or extrapolation of the available concentration data. In this context, interpolation procedures assume that the values of concentration or sediment discharge obtained from instantaneous samples represent a much longer period of time, for example, days or weeks, whereas rating curve techniques may be viewed as the classic example of an extrapolation procedure. In the latter case, a limited number of sediment concentration measurements are extrapolated over the period of interest by developing a relationship between concentration or sediment discharge and stream discharge and applying this relationship to the streamflow record (14, 58, 94). The streamflow record may be in the form of a flow duration curve or a continuous series. Researchers in many studies have applied rating relationships developed from a short period of measurement to streamflow records covering a much longer length of time. Estimates of suspended sediment load obtained using these indirect load calculation procedures may involve considerable error, and the resulting data must be treated with caution.

Figure 3A shows the degree of scatter that may be associated with a suspended sediment concentration/discharge relationship. In this case the concentrations associated with a specific level of discharge range over several orders of magnitude. Although improved relationships between concentration and discharge may be obtained by subdividing the data set according to season and rising and falling stages of the hydrograph, any estimate of sediment concentration derived using such relationships will be associated with a very considerable standard error. Because rating relationships commonly are fitted to logarithmic plots using linear least squares regression and they frequently are biased by low-flow samples, estimates of sediment yield obtained using rating curves can be expected to underestimate the true sediment yield (98), perhaps by as much as 80 percent. For some rivers, scientists cannot assume that the rating relationship remains constant over a period of years and that the available samples can be used to develop a relationship applicable to the entire period because land use change may cause nonstationarity. Dunne (22) reported an interesting example of interannual variation in the sediment discharge/water discharge relationship for the Tana River at Garissa, Kenya (Figure 3B). His data showed an essentially random pattern rather than any progressive shift.

As a final example of potential problems associated with infrequent sampling and use of extrapolation and interpolation techniques to estimate long-term sediment yields, figure 4 compares the actual suspended sediment load of the River Creedy in Devon, England, for a 7-year period with nearly 100 load estimates for the same period; these estimates were obtained using typical (infrequent) manual sampling strategies and a selection of indirect load calculation procedures. In this case, a continuous-recording turbidity meter was used to provide a detailed record of sediment concentration. Walling and Webb (98) used this record to calculate the actual sediment load and as the basis for deriving data sets representing the manual sampling strategies. The various estimates span a wide range, and underestimation by as much as 60 percent is common.

Although published or tabulated sediment yield data may appear authoritative, it is essential that the user carefully considers their reliability and accuracy before using them for further analysis and interpretation. However, few guidelines for assessing reliability and accuracy are available currently. In many cases, acceptance of the

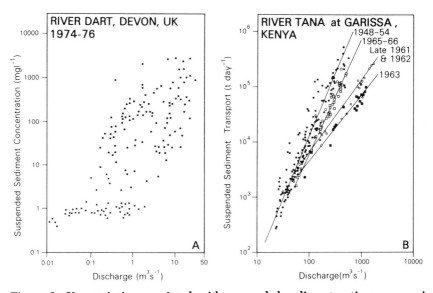

Figure 3. Uncertainties associated with suspended sediment rating curves. A demonstrates the degree of scatter that frequently will be associated with a suspended sediment concentration/discharge relationship, and B illustrates inter-annual variation in the precise form of a suspended sediment load/discharge relationship. Based on Dunne (22).

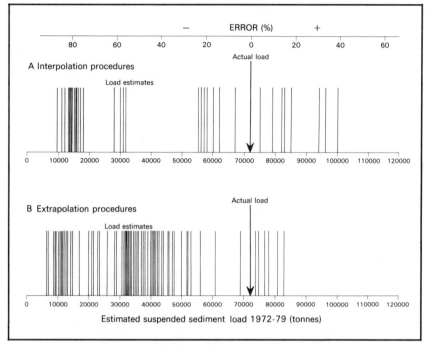

Figure 4. A comparison of suspended sediment load estimates for the River Creedy, Devon, England, obtained using interpolation and extrapolation procedures with the actual load for the period 1972-1979.

limitations of a particular measurement program may permit only very general conclusions about the margins of error that may be involved.

Designing sediment yield monitoring programs

Soil conservation research and planning may require the establishment of a sediment yield monitoring program to supplement or verify existing information or to provide data from a hitherto undocumented area. It is important that the aforementioned problems of data reliability be minimized. For example, if a program is designed to monitor the effects of conservation measures in reducing sediment yields, the sediment load data should be sufficiently accurate to produce meaningful conclusions. Further discussion of monitoring programs can focus on manual sampling strategies, use of automatic sampling

Table 3. Characteristics of U.S.-series depth-integrating and point-integrating samplers for collecting suspended sediment samples.

Name	Type of Sampler*	Method of Suspension	Mass (kg)	Overall Length (m)	Sample Container Size (ml)	Distance Between Nozzle and Sampler Bottom (mm)	Remarks
US DH-48	DI	Rod	2.0	0.33	473	90	For wading
US DH-59	DI	Cable	10.2	.42	473	114	For hand-line operation
US DH-75P	DI	Rod	.4	.26	500	83	For sampling only in subfreezing temperatures
US DH-75Q	DI	Rod	.4	.29	1,000	114	Similar to US DH 75P
US DH-76	DI	Cable	10.9	.47	946	80	Similar to US DH-59
US D-49	DI	Cable	28.0	.61	473	103	
US D-49AL	DI	Cable	18.0	.61	473	103	Similar to US D-49
US D-74	DI	Cable	28.2	.66	473 or 946	103	Similar to US D-49
US D-74AL	DI	Cable	11.4	.66	473 or 946	111	Similar to US D-74
US P-50	PI	Cable	135.6	1.12	473 or 946	140	
US P-61-A1	PI	Cable	47.5	.71	473 or 946	109	
US P-63	PI	Cable	90.4	.86	473 or 946	150	
US P-72	PI	Cable	17.7	.71	473 or 946	109	Similar to US P-61-A1 but for hand-line operation

*Type: DI = depth-integrating, PI = point-integrating.

devices and equipment for recording sediment concentrations, reservoir surveys, and study of sediment properties.

Manual sampling strategies. Several manuals and technical reviews
provide detailed discussions of currently available suspended sediment samplers, sampling techniques, and laboratory procedures for
determining sediment concentration (5, 39, 90, 101, 105). Particular
attention is given to the characteristics of recommended samplers
and to techniques of sample collection. Researchers or policymakers
should consult these publications before initiating a measurement
program. In most circumstances, the sediment samplers developed
by the Federal Interagency Sedimentation Project under the auspices
of the U.S. Water Resources Council (Table 3) will prove appropriate,
and many of these samplers are commercially available from the
Product Manufacturing Company. Figure 5 shows two examples of
suspended sediment samplers used in China.

The manuals give much less attention to sampling frequency
requirements because they are more concerned with measurements
of instantaneous sediment discharge than measurements of long-term
sediment loads and yields. In view of the great uncertainty associated
with any estimate of sediment yield based on infrequent sampling
and the application of indirect load calculation techniques, such as
rating curves, manual sampling programs should aim to document
the continuous record of sediment concentration. This record can
be combined with equivalent water discharge data to provide values
of sediment loads for various time periods.

In designing a sampling strategy to document the continuous record
of sediment concentration, careful consideration must be given to
the likely behavior of the river and the timing of significant concentration fluctuations. During periods of stable flow and low concentration, a low sampling frequency, daily or even weekly, will suffice because concentrations usually remain essentially stable. Any
errors incurred likely would be of limited significance to the final
long-term load estimate. Most of the sediment load will be moved
during major flood events, and it is essential that these are sampled
intensively to record fluctuations in sediment concentration. In the
case of a river that drains a large area, say, greater than 1,000 km^2,
these fluctuations probably will be easy to follow with a sequence
of manual samples, although it may be necessary to sample at frequent intervals, perhaps every 2-3 hours, on the rising stage of the

flood. On smaller rivers, however, fluctuations in sediment concentration may be more rapid, and operational problems may preclude establishing an effective manual sampling program. Alternative automated equipment should be used in these circumstances.

Figure 6 illustrates some of the salient characteristics of suspended sediment transport in the River Creedy, which drains a 262-km^2 basin in Devon, England. The concentration and load-duration curves

Figure 5. These two suspended sediment samplers employed by Chinese scientists on the headwaters of the Yangtze River provide examples of point-integrating (upper) and instantaneous (lower) samplers.

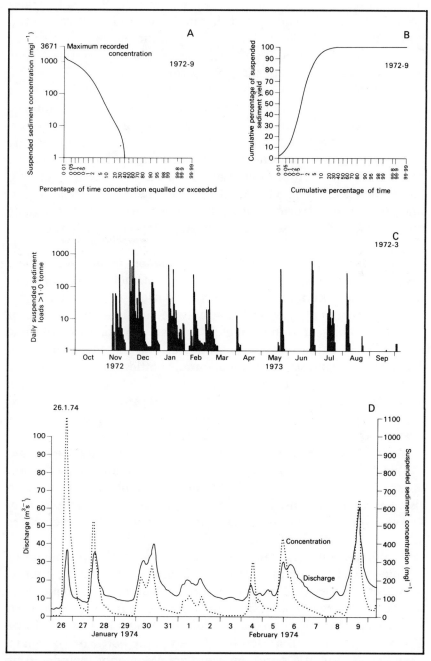

Figure 6. **Characteristics of the suspended sediment regime of the River Creedy, Devon, England.**

(Figure 6A, B) indicate that most of the total sediment load is transported during only 5 percent of the time when sediment concentrations exceed about 100 mg/liter. Clearly, sediment sampling should focus on this time period, although the relatively even distribution of major sediment transporting events through the year (Figure 6C) poses practical problems for undertaking event-based sampling. These problems would be reduced considerably where a river exhibits a marked seasonal regime, with major events being limited to one particular time of year, for example, snowmelt or seasonal rains. Figure 6D demonstrates the impracticality of using extrapolation procedures based on concentration/discharge relationships to reduce the need for frequent sampling. The variation in sediment concentration during a sequence of storm events shows that there is no well-defined relationship between sediment concentration and discharge.

Automatic sampling equipment and continuous monitoring. Where suspended sediment concentrations in a river fluctuate rapidly over time, automatic sampling equipment and devices for continuous monitoring of sediment concentration may be required to augment or replace manual sampling. Single-stage samplers provide a simple means of obtaining samples automatically at an unattended site. These samples consist of individual sample bottles with intake nozzles and exhaust ports (39) mounted in a vertical sequence. The associated concentrations should be viewed only as approximate because (a) samples are collected at a fixed point in the vertical and (b) there may be additional movement of sediment into the bottle during the period that the bottle is submerged. The resulting data are limited further by the fact that samples are obtained at predetermined water levels, rather than on a time basis, and commonly are restricted to rising-stage conditions. During a multipeaked event, samples may be available only from the initial hydrograph peak. However, there are devices that operate during falling stages.

Automatic pumping samplers afford a more effective means of sample collection (Figure 7). These samplers can be programmed to collect samples at predetermined time intervals or increments of stage or discharge. They can be distinguished further according to the two basic principles used for withdrawing samples from the flow: a pump or creation of a vacuum in the sample bottle or a primary chamber. Automatic pumping samplers commonly include 36 or more bottles and in most cases may be battery operated (*3, 90, 96,*

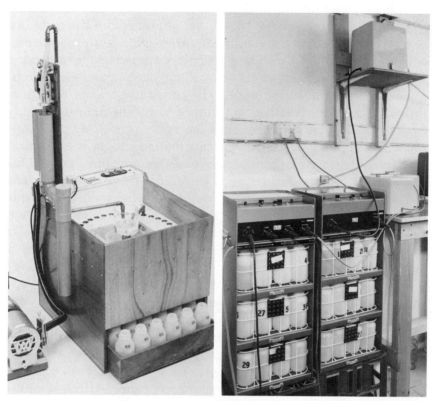

Figure 7. These are typical examples of equipment used for automatic collection of river water samples. Both employ peristaltic pumps to deliver water to the sampler, which incorporates a rotating nozzle capable of filling the set of 36 (left) or 48 (right) bottles.

101). Because the sample is obtained from a fixed inlet and is not collected isokinetically, these samplers are suited more to rivers where the suspended load is comprised predominantly of silt and clay particles. Results obtained should be calibrated against manual samples collected from the overall cross section.

The availability of reliable equipment for continuously monitoring suspended sediment concentrations clearly would be a major benefit in any sediment yield monitoring program, and scientists have devoted considerable effort toward developing such instruments. To date, there is no entirely satisfactory and generally applicable instrument, but several have been used with encouraging results. These include two major types: optical turbidimeters and nuclear gauges.

Optical instruments (*10, 29, 38, 88, 101*) measure the turbidity of water by the attenuation of transmitted light, (absorptiometers) or by monitoring the intensity of the light scattered by the suspended sediment particles (nephelometers). Figure 8 shows an example of a photoelectric turbidity monitor. In both cases, values of optical turbidity, generally defined in terms of formazin standards, must be calibrated against direct measurements of suspended sediment concentration because the particle size distribution and the organic matter content, density, and mineralogy of the sediment particles influences the relationship between turbidity and sediment concentration. With some instruments, fouling of cell windows and ambient light effects also may affect the calibration adversely. In some cases, the sensor may be mounted directly in the river. In others, the river water is circulated by a pump to the measuring cell housed on the river bank.

Reports about the effectiveness of optical turbidity meters for continuous monitoring of suspended sediment concentrations are somewhat conflicting. In some studies, researchers apparently have used turbidity meters successfully. In others, it has been difficult, if not impossible, to develop a meaningful calibration between turbidity and sediment concentration. However, the balance of evidence suggests that such instruments frequently provide an effective means of monitoring sediment concentrations, although they are limited to the range of sediment concentrations below 10,000 mg/liter. Furthermore, they are applicable primarily to the clay and silt fraction of the suspended load. Fish (*27*) demonstrated that optical turbidimeters are virtually insensitive to sand-size material, unless the associated concentrations are high.

Nuclear sediment gauges are based on similar principles but measure the absorption or scattering of gamma radiation rather than light (*7, 52, 74, 84*). They are suitable for concentrations above a minimum threshold of about 500 mg/liter. These gauges commonly use probes that are either fixed within or lowered into the channel. However, the instruments are more sophisticated than optical turbidity meters and have been restricted mostly to specialized experimental measurements rather than routine monitoring. Other devices, such as ultrasonic sensors (*28*) and vibrating U-tube densimeters (*80*), have been used, but these devices are essentially experimental.

Manual sampling and use of equipment for continuous monitor-

Hydraulics Research, Ltd.

Figure 8. A photoelectric turbidity monitor and probe (inset) can be mounted directly in a river, as at this monitoring station on the Sagana River in Kenya, to provide a continuous record of wash load concentrations. The probe is mounted inside the perforated tube for protection and may be readily raised for cleaning.

ing of suspended sediment concentrations frequently may be complementary. For example, continuous-monitoring devices could provide a useful means of interpolating fluctuations in sediment concentrations between manual samples for which accurate values of sediment concentration have been obtained. Bolton (9) suggested that in many instances it may be convenient to distinguish the fine-grained washload and the coarser, suspended bed material components of the total suspended load—using optical turbidity meters to monitor the former and manual sampling to measure the latter. He concluded that it may be possible to obtain a clearly defined rating relationship between suspended bed material concentrations and flow so that the record for this component can be extrapolated from limited sampling.

Reservoir surveys. Reservoir sedimentation surveys assess the sediment yield from drainage basins that discharge into reservoirs (*17*). Such surveys have several advantages over river measurements. For example, a single survey in many cases will provide a meaningful estimate of sediment yield averaged over a period of several years. Provided the trap efficiency of the reservoir can be estimated and the volumes and densities of deposited sediment assessed accurately, there should be less uncertainty associated with the resulting sediment yield values. Furthermore, once the ranges and benchmarks are established, resurveys can be undertaken relatively easily. In addition to an estimate of the trap efficiency of the reservoir, information on the history of the reservoir, including the date of impoundment and whether any sediment has been removed, should be available. Sediment yields calculated from reservoir survey data, however, will include both bedload and suspended load.

Procedures for reservoir sedimentation surveys have been developed for both small and large impoundments (5, 75, 76, 90). Sonic depth recorders measure water depth along predetermined ranges, and sediment depth can be calculated from a previous survey of the original reservoir bottom or from soundings of sediment depth. In view of the difficulties of sounding sediment depth, it is advisable to undertake an initial survey of the bottom before or immediately after impoundment. Values of volume-weight or density of deposited sediments can be obtained by collecting undisturbed volumetric samples of the deposits or by in-situ measurements using a gamma probe (55). Use of the latter, however, is restricted to saturated sediments.

To determine the mean annual sediment yield from the contributing drainage basin, the weight of deposited sediment must be adjusted for reservoir sediment trap efficiency. Heinemann (43) discussed some of the problems involved in obtaining accurate estimates of trap efficiency. Clearly, there is a need for further research on this topic because most existing estimation procedures are based on data assembled from reservoirs in the United States. In the absence of more detailed procedures, many researchers use the curves developed by Brune (13), which relate trap efficiency to the capacity/inflow ratio of the reservoir and an approximate index of sediment particle size.

Study of sediment properties. Traditionally, studies of suspended sediment transport by rivers have emphasized documentation of the magnitude of the loads transported. There now is increased awareness of sediment's role in the transport of nutrients and contaminants and in nonpoint pollution from agricultural areas (34, 81, 89). This has focused attention on the properties of the sediment transported—either the physical and chemical properties of the sediment as a component of river water quality or the relationship between the properties of sediment and source material and, therefore, the selectivity of the erosion and conveyance processes. The latter is perhaps more relevant to soil erosion and conservation because erosion by water may preferentially mobilize and transport specific fractions and constituents of the soil that are of particular importance in maintaining fertility. Therefore, attention may be directed to particle size characteristics and the organic matter, nutrient, and mineral content of transported sediment and associated enrichment ratios (100). For example, figure 9 illustrates the potential contrasts in the particle size of suspended sediment and source material. It compares the proportion of clay, silt, and sand in suspended sediment samples collected by the U.S. Geological Survey from four small drainage basins with equivalent information on the soils, based on their textural classification. Table 4, based on my work in England, compares a number of suspended sediment properties with equivalent data for soils in the drainage basins.

In some cases, laboratories can analyze properties of the small samples of particulate material recovered from manual or automatic samplers. However, where a wide range of analysis is intended or where a particular laboratory procedure requires a considerable

amount of sediment, researchers must collect bulk samples of suspended sediment. Experience suggests that continuous flow centrifuges are most useful in recovering sediment from bulk samples of
river water. Ongley and Blachford (67) reported the use of a specialized sampling/recovery system in which water is pumped directly
from the river into the centrifuge. They found good recovery performance for particles down to 0.45 μm.

Lake sediment studies. The lack of long-term sediment transport
records hampers the study of long-term patterns of sediment yield
from a drainage basin. The availability of records covering, for

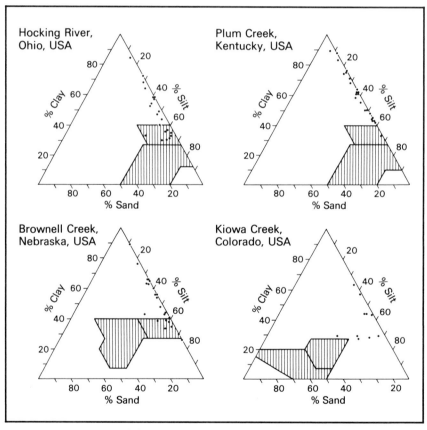

Figure 9. Comparison of the particle size composition of suspended sediment and
soils in four U.S. drainage basins. Textural classes of the dominant soils are denoted
by shaded zones. Based on data from Flint (30), Anttila (6), and Mundorff (61, 62).

Table 4. Comparison of the properties of suspended sediment and source-area material for several rivers in Devon, England.

	Total Percent			Iron (%)			
	C	N	P	Fed*	Feox†	χ‡	ΣC§
Jackmoor Brook							
Suspended sediment	4.84	0.58	0.147	2.48	0.74	7.60	28
Soils							
Arable (slopes)	2.61	0.29	0.123	2.01	0.39	4.03	14
Pasture (slopes)	3.91	0.39	0.126	2.47	0.54	9.15	13
Floodplain pasture	3.91	0.28	0.104	2.25	0.52	6.01	12
River Dart							
Suspended sediment	6.61	0.64	0.148	2.02	1.06	3.88	24
Soils							
Arable (slopes)	1.48	0.16	0.081	1.68	0.27	4.45	10
Pasture (slopes)	5.29	0.50	0.159	1.75	0.57	4.63	16
Alluvial pasture very poorly represented in this catchment.							
River Exe							
Suspended sediment	7.11	0.68	0.153	2.07	0.95	4.33	28
Soils							
Arable (slopes)	2.16	0.22	0.120	1.47	0.37	6.77	16
Pasture (slopes)	4.63	0.44	0.144	1.71	0.53	5.70	14
Floodplain pasture	4.50	0.45	0.105	1.87	0.56	-	8
Banks	<1.0	<0.1	<0.05	-	-	2	1-2

*Total 'free' iron (dithionite extraction).
†Organic and inorganic 'amorphous' iron (oxalate extraction).
‡Magnetic susceptibility m^3g^{-1}
§Exchangeable cations meq $100g^{-1}$

example, the last 100 years would be of considerable value in attempting to understand temporal variations in sediment yield in response to changes in catchment condition. Record lengths increase as time proceeds, but opportunities to document particular changes in sediment response may not recur. In this context, researchers should focus more attention on the potential of lake sedimentation studies in illuminating the environmental history of the upstream drainage basin (64, 69). The work of Davis (15) in Frains Lake, Michigan, in the United States provides a classic example of the potential of such work. The lake is located at the outlet of a 0.18-km² drainage basin. With detailed sediment coring and core analysis and dating, Davis reconstructed the pattern of inorganic sediment influx and sediment yield over the previous 200 years (Figure 10A). The data show low rates of sediment yield in presettlement times, rising by a factor of up to 70 with the onset of settlement and agricultural clearance after 1830,

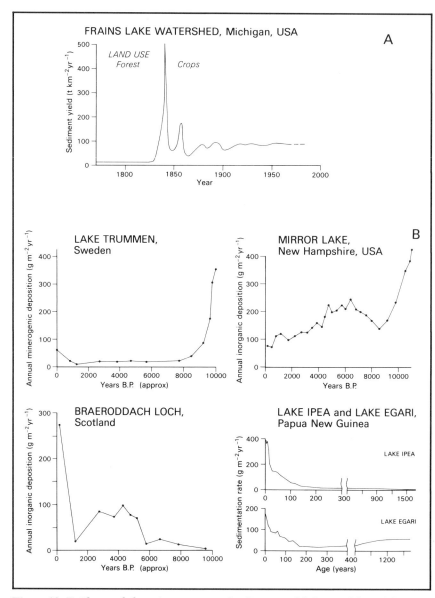

Figure 10. Evidence of changing patterns of sediment yield obtained from lake sediment studies. This illustrates the reconstruction of (A) sediment yields to Frains Lake, Michigan, proposed by Davis (15) and of (B) sedimentation rates in Mirror Lake, New Hampshire; Lake Trummen, Sweden; Braeroddach Loch, Scotland; and two small lakes in Papua, New Guinea, based on data presented by Likens and Davis (50), Digerfeldt (20), Edwards and Rowntree (25), and Oldfield and associates (65), respectively.

and stabilizing after 1900 at a rate about 10 times the presettlement rate.

Davis' work at Frains Lake necessitated analysis of a considerable number of cores to calculate the volumes of inorganic sediment involved. Estimates of sediment yield from the catchment area require this volume data. In many instances, it may be impossible to collect more than a small number of cores, and changes in sediment yield over the period represented by the core must be inferred from changes in the rate of sediment accumulation at those points. The researchers then must assume that the pattern of sedimentation over the lake floor has remained constant through time—an assumption that may be unjustified. Data on changes in sediment yield over time, based on analysis of single cores, must, therefore, be treated cautiously, but the information may nevertheless prove useful. Figure 10B provides four examples of such studies. In this case, the vertical axes of the plots refer to sediment accumulation rates rather than to sediment yield. The core data from Papua, New Guinea, show the impact of human activity on sedimentation rates during the past 300 years. Oldfield and associates (65) tentatively related the increased sedimentation to the intensification of land use resulting from the introduction of the sweet potato and, more definitively, to the post-1950 impact of "Western" peoples.

Development of techniques for rapid intercore correlation using magnetic measurements (8) could facilitate the calculation of sediment volumes and, therefore, sediment yields from lake sediment studies. Interpretational problems related to defining the trap efficiency of the lake body, dating the sediment cores, and distinguishing allochthonous and autochthonous sources of minerogenic particulate material still remain. But this general approach merits further exploitation where suitable lakes exist. Detailed analysis of the stratigraphic distribution of sediment characteristics in both lakes and reservoirs also may provide additional information on changes in sediment sources and erosional history (102).

Linking downstream sediment to upstream erosion

As mentioned, there are several problems involved in linking downstream sediment yields to upstream rates of soil erosion. This task faces many uncertainties. The traditional approach in theory involves the application of a SDR to calculate the gross erosion within the drainage basin, that is:

Gross erosion = sediment yield/SDR

In practice, however, few rigorous guidelines exist for predicting the SDR for a specific basin (95).

The magnitude of the SDR for a particular basin is influenced by a wide range of geomorphological and environmental factors, including the nature, extent, and location of the sediment sources; relief

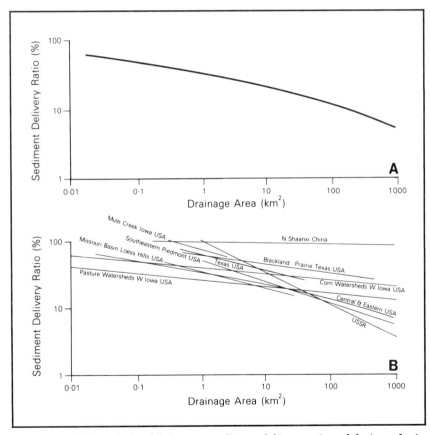

Figure 11. (A) The relationship between sediment delivery ratio and drainage basin area developed by the U.S. Soil Conservation Service for the central and eastern United States and (B) similar relationships proposed by other workers for different areas. B is based on Roehl (79), Sokolovskii (82), Piest and associates (72), Renfro (77), ASCE (5), Williams (103), and Mou and Meng (60).

Table 5. Examples of proposed relationships between sediment delivery ratio and catchment characteristics.

Region	Equation*	Author
Kansas, USA	$\log DR = 2.962 + 0.869 \log R - 0.854 \log L$	(53)
Southeastern USA	$\log DR = 4.5 - 0.23 \log 10\ A - 0.510 \times \text{colog } R/L - 2.786 \log BR$	(79)
Brushy Creek, Texas, USA	$DR = 0.627\ SLP^{0.403}$	(104)
Texas, USA	$DR = 1.366 \times 10^{-11}\ A^{-0.100}\ R/L^{0.363}\ CN^{5.444}$	(103)
Pigeon Roost Creek, Mississippi, USA	$DR = 0.488 - 0.006\ A + 0.010\ RO$	(63)
Dali River Basin, Shaanxi, China	$DR = 1.29 + 1.37 \ln Rc - 0.025 \ln A$	(60)

*DR = sediment delivery ratio; R = basin relief; L = basin length; A = basin area; R/L = relief/length ratio; BR = bifurcation ratio; SLP = percent slope of main stem channel; CN = SCS curve number; RO = annual runoff; Rc = gully density. Note: units vary between equations.

and slope characteristics; the drainage pattern and channel conditions; vegetation cover; land use; and soil texture. Several studies have attempted to produce empirical prediction equations for this variable. Basin area frequently has been isolated as a dominant control (5). The U.S. Soil Conservation Service has developed a generalized relationship between SDR and drainage basin area (Figure 11A), which has been widely applied in the central and eastern United States (79). The inverse relationship between SDR and area has been explained in terms of decreasing slope and channel gradients and increasing opportunities for deposition associated with increasing basin size. Other attempts to develop prediction equations have included such variables as relief ratio and main channel slope (Table 5).

The failure to produce a generally applicable prediction equation for SDR is due partly to the complexity of sediment delivery processes and their interaction with catchment characteristics and partly to a lack of definitive assessments of the dependent variable. Assessments that have been undertaken are themselves based primarily on a comparison of measured sediment yield with an estimate of gross erosion. The latter is generally derived from an estimate of sheet erosion based on a soil loss equation, corrected to take account of additional contributions from channel and gully erosion. These estimates of gross erosion, therefore, are open to considerable uncertainty. The

various relationships developed for particular areas presented in figure
11B emphasize the considerable diversity of values associated with
basins of a given size. The delivery characteristics of basins in the
southeastern Piedmont in the United States contrast markedly with
those of watersheds in the loess region of China, where delivery ratios
close to 100 percent occur over a wide range of basin size. Some con-

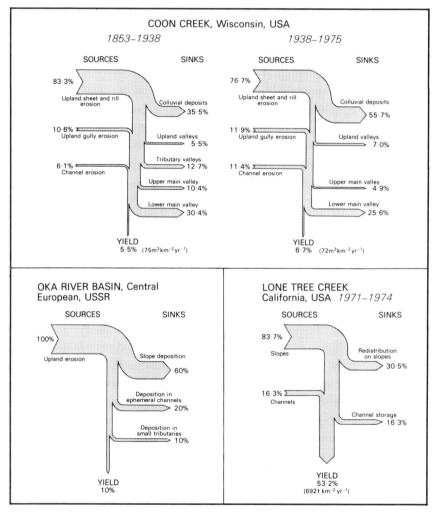

Figure 12. Tentative sediment budgets for Coon Creek, Wisconsin (360km²),
1853-1938 and 1938-1975; Lone Tree Creek, California (1.74 km²); and the Oka
River, USSR. Based on data presented by Trimble (87), Lehre (48), and Zavlasky
(106).

sistency in the slope of the delivery ratio-area relationship is apparent for several regions in the United States, lending support to the view that an exponent of -0.125 is typical (5). But even here a wide range of intercept values exists.

A more rigorous approach to defining and investigating the sediment delivery characteristics of a drainage basin is provided by the sediment budget concept originally advocated by Dietrich and Dunne (19) and developed by Lehre (47, 48) and Swanson and associates (83). In this essentially conceptual approach, the various sediment sources within a basin are defined, and the sediment mobilized from these sources is routed to and through the channel system by considering the various sinks. Figure 12 provides diagrammatic representations of such budgets for three basins where investigators have attempted to define tentatively the source-yield linkages. In all cases, the proportion of eroded sediment delivered to the basin outlet is relatively small, ranging from 53 percent to 5.5 percent, but considerable differences exist between the catchments represented in the precise form of the budget and in the location and importance of the various sinks.

As yet, the availability of techniques for quantifying the various sources and sinks involved in a sediment budget is extremely limited. Development of such techniques represents an important research need. The use of sediment properties to fingerprint sediment sources (46, 66, 93, 100) and the use of ^{137}Cs measurements to investigate the amount and spatial distribution of deposited sediment within a drainage basin (12, 51, 54, 56) must, however, be seen as important advances toward this goal.

REFERENCES

1. Acres/ILACO. 1965. *A survey of the irrigation potential of the Lower Tana River.*
2. Adams, J. 1980. *High sediment yields from major rivers of the western Southern Alps, New Zealand.* Nature 287: 88-89.
3. Allen, P. B. 1981. *Measurement and prediction of erosion and sediment yield.* ARM-S-15. U.S. Department of Agriculture, Washington, D.C.
4. Allen, P. B., and D. V. Petersen. 1981. *A study of the variability of suspended sediment measurements.* In *Erosion and Sediment Transport Measurement, Proceedings, Florence Sympoisum, June 1981.* Publication No. 133. International Association of Hydrological Sciences, Wallingford, England. pp. 203-211.
5. American Society of Civil Engineers. 1975. *Sedimentation engineering.* New York, New York. 745 pp.
6. Anttila, P. W. 1970. *Sedimentation in Plum Creek subwatershed No. 4, Shelby*

County, North-Central Kentucky. Water Supply Paper 1798-G. U.S. Geological Survey, Reston, Virginia.

7. Berke, B., and L. Rakoczi. 1981. *Latest achievements in the development of nuclear suspended sediment gauges.* In *Erosion and Sediment Transport Measurement, Proceedings, Florence Symposium, June 1981.* Publication No. 133. International Association of Hydrological Sciences, Wallingford, England, pp. 91-96.

8. Bloemendal, J., F. Oldfield, and R. Thompson. 1979. *Magnetic measurements used to assess sediment influx at Llyn Goddionduon.* Nature 280: 50-53.

9. Bolton, P. 1983. *Sediment discharge measurement and calculation.* Technical Note OD/TN2. Hydraulics Research, Wallingford, England.

10. Brabben, T. 1981. *Use of turbidity monitors to assess sediment yield in East Java, Indonesia.* In *Erosion and Sediment Transport Measurement, Proceedings, Florence Symposium, June 1981.* Publication No. 133. International Association of Hydrological Sciences, Wallingford, England, pp. 105-113.

11. Branski, J. 1975. *Ocena denudacji dorzecze wisley na podatawie wynikow ponmiarow rumowiska unoszonego.* Prace Instytyta Meteorologii i Gospodarki Wodnej 6: 1-58.

12. Brown, R. B., G. F. Kling, and N. H. Cutshall. 1981. *Agricultural erosion indicated by ^{137}Cs redistribution: II. Estimates of erosion rates.* Soil Science Society of America Journal 45: 1,191-1,197.

13. Brune, G. M. 1953. *Trap efficiency of reservoirs.* Transactions, American Geophysical Union 34: 407-418.

14. Campbell, F. B., and H. A. Bauder. 1940. *A rating-curve method for determining silt-discharge of streams.* Transactions, American Geophysical Union 21: 603-607.

15. Davis, M. B. 1976. *Erosion rates and land use history in southern Michigan.* Environmental Conservation 3: 139-148.

16. Demmak, A. 1982. *Contribution a létude de lérosion et des transports solides en Algerie Septentrionale.* These de Docteur-Ingenieur, Universite de Paris, Paris, France.

17. Dendy, F. E., and G. C. Bolton. 1976. *Sediment yield-runoff drainage area relationships in the United States.* Journal of Soil and Water Conservation 32: 264-266.

18. Dickinson, W. T. 1981. *Accuracy and precision of suspended sediment loads.* In *Erosion and Sediment Transport Measurement, Proceedings, Florence Symposium, June 1981.* Publication No. 133. International Association of Hydrological Sciences, Wallingford, England. pp. 195-202.

19. Dietrich, W. B., and T. Dunne. 1978. *Sediment budget for a small catchment in mountainous terrain.* Zeitschrift fur Geomorphologie, Supplementband. 29: 191-206.

20. Digerfeldt, G. 1972. *The post-glacial development of Lake Trummen.* Folia Limnologica Scandinavica 16.

21. Douglas, I. 1973. *Rates of denudation in selected small catchments in eastern Australia.* Occasional Papers in Geography No. 21. University of Hull, Hull, England.

22. Dunne, T. 1977. *Evaluation of erosion conditions and trends.* In *Guidelines for Watershed Management.* Conservation Guide No. 1. Food and Agriculture Organization, United Nations, Rome, Italy. pp. 53-83.

23. Dunne, T., and G.S.O. Ongweny. 1976. *A new estimate of sedimentation rates on the Upper Tana River.* Kenyan Geographer 2: 20-38.

24. Edwards, K. A. 1979. *Regional contrasts in rates of soil erosion and their significance with respect to agricultural development in Kenya.* In R. Lal and D. J. Greenland [editors] *Soil Physical Properties and Crop Production in the Tropics.* John Wiley & Sons, Chichester, England. pp. 441-454.
25. Edwards, K. J., and K. M. Rowntree. 1980. *Radiocarbon and palaeo-environmental evidence for changing rates of erosion at a Flandrian stage site in Scotland.* In R. A. Cullingford, D. A. Davidson, and J. Lewin [editors] *Timescales in Geomorphology.* John Wiley & Sons, Chichester, England. pp. 207-223.
26. Emmett, W. W. 1984. *Measurement of bedload in rivers.* In R. F. Hadley and D. E. Walling [editors] *Erosion and Sediment Yield: Some Methods of Measurement and Modelling.* Geobooks, Norwich, England. pp. 91-109.
27. Fish, I. L. 1983. *Partech turbidity monitors. Calibration with silt and the effects of sand.* Technical Note OD/TNR. Hydraulics Research, Wallingford, England.
28. Flammer, G. H. 1962. *Ultrasonic measurement of suspended sediment.* Bulletin 1141-A. U.S. Geological Survey, Reston, Virginia.
29. Fleming, G. 1969. *Suspended solids monitoring: A comparison between three instruments.* Water and Waste Engineering 72: 377-382.
30. Flint, R. F. 1972. *Fluvial sediment in Hocking River subwatershed 1 (North Branch Hunters Run) Ohio.* Water Supply Paper 1798-I. U.S. Geological Survey, Reston, Virginia.
31. Food and Agriculture Organization, United Nations. 1979. *A provisional methodology for soil degradation assessment.* Rome, Italy.
32. A. Gibb and Partners. 1959. *Upper Tana water resources survey 1958-1959.* Reading, England.
33. Glymph, L. M. 1954. *Studies of sediment yields from watersheds.* Publication No. 36. International Association of Hydrological Sciences, Wallingford, England. pp. 173-191.
34. Golterman, H. L., P. G. Sly, and R. L. Thomas. 1983. *Study of the relationship between water quality and sediment transport.* Technical Papers in Hydrology No. 26. UNESCO, Paris, France.
35. Griffiths, G. A. 1979. *High sediment yields from major rivers of the western Southern Alps, New Zealand.* Nature 282: 61-63.
36. Griffiths, G. A. 1981. *Some suspended sediment yields from South Island catchments, New Zealand.* Water Resources Bulletin 17: 662-671.
37. Griffiths, G. A. 1982. *Spatial and temporal variability in suspended sediment yields of North Island basins, New Zealand.* Water Resources Bulletin 18: 575-584.
38. Grobler, D. C., and A. van B. Weaver. 1981. *Continuous measurement of suspended sediment in rivers by means of a double beam turbidity meter.* In *Erosion and Sediment Transport Measurement, Proceedings, Florence Symposium, June 1981.* Publication No. 138. International Association of Hydrological Sciences, Wallingford, England. pp. 97-103.
39. Guy, H. P., and V. W. Norman. *Field methods for measurement of fluvial sediment.* In *Techniques of Water Resources Investigations of the U.S. Geological Survey* (Book 5, Chapter C1). U.S. Geological Survey, Reston, Virginia.
40. Hadley, R. F., and D. E. Walling, editors. *Erosion and sediment yield: Some methods of measurement and modelling.* Geobooks, Norwich, England.
41. Hadley, R. F., and L. M. Shown. 1976. *Relation of erosion to sediment yield.* In *Proceedings, Third Federal Inter-Agency Sedimentation Conference.* U.S.

Water Resources Council, Washington, D.C. 1-132—1-139.
42. Hardjowitjitro, H. 1981. *Soil erosion as a result of upland traditional cultivation in Java Island.* In T. Tingsanchali and H. Eggars [editors] *Proceedings, South-East Asia Regional Symposium on Problems of Soil Erosion and Sedimentation.* Asian Institute of Technology, Bangkok, Thailand. pp. 173-179.
43. Heinemann, H. G. 1984. *Reservoir trap efficiency.* In R. F. Hadley and D. E. Walling [editors] *Erosion and Sediment Yield: Some Methods of Measurement and Modelling.* Geobooks, Norwich, England. pp. 201-218.
44. Heusch, B. 1980. *Erosion in the Ader Dutchi Massif (Niger).* In M. De Boodt and D. Gabriels [editors] *Assessment of Erosion.* John Wiley & Sons, Chichester, England. pp. 521-529.
45. International Standards Organization. 1976. *Methods for measurement of suspended sediment.* Draft International Standard, 4363. New Delhi, India.
46. Klages, M. G., and Y. P. Hsieh. 1975. *Suspended solids carried by the Gallatin River of southwestern Montana: II. Using mineralogy for inferring sources.* Journal of Environmental Quality 4: 68-73.
47. Lehre, A. K. 1981. *Sediment budget of a small coast range drainage basin.* In *Erosion and Sediment Transport in Pacific Rim Steeplands.* Publication No. 132. International Association of Hydrological Sciences, Wallingford, England. pp. 123-139.
48. Lehre, A. K. 1982. *Sediment budget in a small coast range drainage basin in north-central California.* In F. J. Swanson, R. J. Janda, T. Dunne, and D. N. Swanston [editors] *Sediment Budgets and Routing in Forested Drainage Basin.* General Technical Report PNW-141. Forest Service, U.S. Department of Agriculture, Portland, Oregon.
49. Li, Y-H. 1976. *Denudation of Taiwan Island since the Pliocene epoch.* Geology 4: 105-107.
50. Likens, G. E., and M. B. Davis. 1975. *Postglacial history of Mirror Lake and its watershed in New Hampshire, USA: An initial report.* Verhalt International Verein Limnologie 19: 982-993.
51. Loughran, R. J., B. L. Campbell, and G. L. Elliott. 1982. *The identification and quantification of sediment sources using* [137]*Cs.* In D. E. Walling [editor] *Recent Developments in the Explanation and Prediction of Erosion and Sediment Yield.* Publication No. 137. International Association of Hydrological Sciences, Wallingford, England. pp. 361-369.
52. Lu Zhi, Yuren Lui, Leling Sun, Xianglin Xu, Yujing Yang, and Lingqui Kong. 1981. *The development of nuclear sediment concentraton gauges for use on the Yellow River.* In *Erosion and Sediment Transport Measurement, Proceedings, Florence Symposium.* Publication No. 133. International Association of Hydrological Sciences, Wallingford, England. pp. 83-90.
53. Maner, S. B. 1958. *Factors affecting sediment delivery rates in the Red Hills physiographic area.* Transactions, American Geophysical Union 39: 669-675.
54. McCallan, M. E., B. M. O'Leary, and C. W. Rose. 1980. *Redistribution of cesium-137 by erosion and deposition on Australian soil.* Australian Journal of Soil Research 18: 119-128.
55. McHenry, J. R., and F. E. Dendy. 1964. *Measurement of sediment density by attenuation of transmitted gamma rays.* Soil Science Society of America Proceedings 28: 817-822.
56. McHenry, J. R., and J. C. Ritchie. 1977. *Estimating field erosion losses from fall out cesium-137 measurements.* In *Erosion and Solid Matter Transport in Inland Waters, Proceedings, Paris Symposium, July 1977.* Publication 122.

International Association of Hydrological Sciences, Wallingford, England. pp. 26-33.

57. Meade, B. H. 1982. *Sources, sinks and storage of river sediment in the Atlantic drainage of the United States.* Journal of Geology. 90: 235-252.
58. Miller, C. R. 1951. *Analysis of flow duration sediment rating curve method of computing sediment yield.* U.S. Bureau of Reclamation. Washington, D.C.
59. Moorhead, H. J., and G. P. Sims. 1982. *Sediment deposition in reservoirs on the River Tana, Kenya.* Proceedings, 14th Congress on Large Dams, Rio de Janeiro, Brazil. pp. 601-613.
60. Mou, J., and Q. Meng. 1980. *Sediment delivery ratio as used in the computation of the watershed sediment yield.* Beijing, China.
61. Mundorff, J. C. 1964. *Fluvial sediment in Kiowa Creek basin, Colorado.* Water Supply Paper 1798-A. U.S. Geological Survey, Reston, Virginia.
62. Mundorff, J. C. 1966. *Sedimentation in Brownell Creek, Subwatershed No. 1, Nebraska.* Water Supply Paper 1798-C. U.S. Geological Survey, Reston, Virginia.
63. Mutchler, C. K., and A. J. Bowie. 1976. *Effect of land use on sediment delivery ratios.* In *Proceedings, Third Federal Inter-Agency Sedimentation Conference.* U.S Water Resources Council, Washington D.C. 1-11-1-21.
64. Oldfield, F. 1977. *Lakes and their drainage basins as units of sediment-based ecological study.* Progress in Physical Geography 1: 460-504.
65. Oldfield, F., P. G. Appleby, and P. Thompson. 1980. *Palaeoecological studies of lakes in the Highlands of Papua, New Guinea. I. The chronology of sedimentation.* Journal of Ecology 68: 457-477.
66. Oldfield, F., T. A. Rummery, R. Thompson, and D. E. Walling. 1979. *Identification of suspended sediment sources by means of magnetic measurements: Some preliminary results.* Water Resources Research. 15: 211-218.
67. Ongley, B. D., and D. P. Blachford. 1982. *Application of continuous-flow centrifugation to contaminant analysis of suspended sediment in fluvial streams.* Environmental Technology Letters 3: 219.
68. Ongweny, G.S.O. 1978. *Erosion and sediment transport in the Upper Tana catchment.* Ph.D. thesis. University of Nairobi, Nairobi, Kenya.
69. O'Sullivan, P. E. 1979. *The ecosystem watershed concept in the environmental sciences—a review.* International Journal of Environmental Studies 13: 273-281.
70. Oyebande, L. 1981. *Sediment transport and river basin management in Nigeria.* In R. Lal and E. W. Russell [editors] *Tropical Agricultural Hydrology.* John Wiley & Sons, Chichester, England. pp. 201-225.
71. Pickup, G., R. J. Higgins, and R. F. Warner. 1981. *Erosion and sediment yield in Fly River drainage basins.* In *Erosion and Sediment Transport in Pacific Rim Steeplands.* Publication No. 132. International Association of Hydrological Sciences, Wallingford, England. pp. 438-456.
72. Piest, R. F., L. A. Kramer, and H. G. Heinemann. 1975. *Sediment movement from loessial watersheds.* In *Present and Prospective Technology for Predicting Sediment Yields and Sources.* ARS-S-40. Agricultural Research Service, U.S. Department of Agriculture, Washington, D.C. pp. 130-141.
73. Porterfield, G. 1972. *Computation of fluvial-sediment discharge.* In *Techniques of Water Resources Investigations of the U.S. Geological Survey* (Book 3, Chapter C3). U.S. Geological Survey, Reston, Virginia.
74. Rakoczi, L. 1976. *Possibilities and limitations of nuclear suspended sediment gauging.* In *Modern Developments in Hydrometry.* Publication No. 427. World

Meterological Organization, Geneva, Switzerland. pp. 138-150.

75. Rausch, D. L., and H. G. Heinemann. 1976. *Reservoir sedimentation survey methods.* In *Hydrological Techniques for Upstream Conservation.* Conservation Guide No. 2. Food and Agriculture Organization, United Nations, Rome, Italy. pp. 29-43.

76. Rausch, D. L., and H. G. Heinemann. 1984. *Measurement of reservoir sedimentation.* In R. F. Hadley and D. E. Walling [editors] *Erosion and Sediment Yield: Some Methods of Measurement and Modelling.* Geobooks, Norwich, England. pp. 179-200.

77. Renfro, G. W. 1975. *Use of erosion equations and sediment delivery ratios for predicting sediment yield.* In *Present and Prospective Technology for Predicting Sediment Yields and Sources.* ARS-S-40. Agricultural Research Service, U.S. Department of Agriculture, Washington, D.C. pp. 33-45.

78. Robinson, A. R. 1977. *Relationships between soil erosion and sediment delivery.* In *Erosion and Solid Matter Transport in Inland Waters, Proceedings, Paris Symposium, July 1977.* Publication No. 122. International Association of Hydrological Sciences, Wallingford, England. pp. 159-167.

79. Roehl, J. E. 1962. *Sediment source areas, delivery ratios and influencing morphological factors.* Publication 59. International Association of Hydrological Sciences, Wallingford, England. pp. 202-213.

80. Seely, E. H. 1982. *The Goodwin Creek Research Catchment: Part IV. Field Data Acquisition.* In *Hydrological Research Basins and Their Use in Water Resources Planning, Proceedings, Berne Symposium, September 1982.* Landeshydrologie, Berne, Switzerland. pp. 173-182.

81. Shear, H., and A.E.P. Watson, editors. 1977. *The fluvial transport of sediment-associated nutrients and contaminants.* International Joint Commission, Windsor, Ontario.

82. Sokolovskii, D. L. 1968. *Rechnoi Stok. Osnovy Teorii i Metodiki Paschetov.* Gidrometeorologicheskoe Izdatel'stvo, Leningrad, Soviet Union.

83. Swanson, F. J., R. J. Janda, T. Dunne, and D. N. Swanston. 1982. *Sediment budgets and routing in forested drainage basins.* General Technical Report PNW-141. Forest Service, U.S. Department of Agriculture, Portland, Oregon.

84. Tazioli, G. S. 1981. *Nuclear techniques for measuring sediment transport in natural streams—examples from instrumented basins.* In *Erosion and Sediment Transport Measurement, Proceedings, Florence Symposium, June 1981.* Publication No. 133. International Association of Hydrological Sciences, Wallingford, England. pp. 63-81.

85. Trimble, S. W. 1974. *Man-induced soil erosion on the Southern Piedmont.* Soil Conservation Society of America, Ankeny, Iowa.

86. Trimble, S. W. 1976. *Sedimentation in Coon Creek Valley, Wisconsin.* In *Proceedings, Third Federal Inter-Agency Sedimentation Conference.* U.S. Water Resources Council, Washington, D.C. pp. 5-100-5-112.

87. Trimble, S. W. 1981. *Changes in sediment storage in the Coon Creek Basin, Driftless Area, Wisconsin, 1853 to 1975.* Science 214: 181-183.

88. Truhlar J. F. 1978. *Determining suspended sediment loads from turbidity records.* Hydrological Sciences Bulletin 23: 409-417.

89. U.S. Department of Agriculture. 1976. *Control of water pollution from cropland. Volume II. An overview.* Washington, D.C.

90. U.S. Geological Survey. 1978. *Sediment.* In *National Handbook of Recommended Methods for Water-Data Acquisition.* Reston, Virginia.

91. Virgo, K. J., and R. N. Munro. 1978. *Soil and erosion features of the Central*

Plateau region of Tigrai, Ethiopia. Geoderma 20: 131-157.
92. Wade, J. C., and E. O. Heady. 1978. *Measurement of sediment control impacts on agriculture.* Water Resources Research 14: 1-8.
93. Wall, C. J., and L. P. Wilding. 1976. *Mineralogy and related parameters of fluvial suspended sediments in northwestern Ohio.* Journal of Environmental Quality 5: 168-173.
94. Walling, D. E. 1977. *Limitations of the rating curve technique for estimating suspended sediment loads, with particular reference to British rivers.* In *Erosion and Solid Matter Transport in Inland Waters, Proceedings, Paris Symposium, July 1977.* Publication No. 122. International Association of Hydrological Sciences, Wallingford, England. pp. 34-48.
95. Walling D. E. 1983. *The sediment delivery problem.* Journal of Hydrology 65: 209-237.
96. Walling, D. E. 1984. *Dissolved loads and their measurement.* In R. F. Hadley and D. E. Walling [editors] *Erosion and Sediment Yield: Some Methods of Measurement and Modelling.* Geobooks, Norwich, England. pp. 111-177.
97. Walling, D. E., and A. Teed. 1971. *A simple pumping sampler for research into suspended sediment transport in small catchments.* Journal of Hydrology 13: 325-337.
98. Walling, D. E., and B. W. Webb. 1981. *The reliability of suspended sediment load data.* In *Erosion and Sediment Transport Measurement, Proceedings, Florence Symposium, June 1981.* Publication No. 133. International Association of Hydrological Sciences, Wallingford, England. pp. 177-194.
99. Walling, D. E., and B. W. Webb. 1983. *Patterns of sediment yield.* In K. J. Gregory [editor] *Background to Palaeohydrology.* John Wiley & Sons, Chichester, England. pp. 69-100.
100. Walling, D. E., and P. Kane. 1984. *Suspended sediment properties and their geomorphological significance.* In T. P. Burt and D. E. Walling [editors] *Catchment Experiments in Fluvial Geomorphology.* Geobooks, Norwich, England. pp. 311-334.
101. Ward, P.R.B. 1984. *Measurement of sediment yields.* In R. F. Hadley and D. E. Walling [editors] *Erosion and Sediment Yield: Some Methods of Measurement and Modelling.* Geobooks, Norwich, England. pp. 67-70.
102. Wasson, R. J., R. L. Clark, I. R. Willett, J. Waters, B. L. Campbell, and D. Outhet. 1984. *Erosion history from sedimentation in Burrinjack reservoir, NSW.* In *Drainage Basin Erosion and Sedimentation, Proceedings, Newcastle, NSW, Conference, May 1984.* Soil Conservation Service, New South Wales, Australia. pp. 221-229.
103. Williams, J. R. 1977. *Sediment delivery ratios determined with sediment and runoff models.* In *Erosion and Solid Matter Transport in Inland Waters, Proceedings, Paris Symposium, July 1977.* Publication No. 122. International Association of Hydrological Sciences, Wallingford, England. pp. 168-179.
104. Williams, J. R., and H. D. Berndt. 1972. *Sediment yield computed with universal equation.* Proceedings, American Society of Civil Engineers, Journal of Hydraulics Division 98(HY2): 2,087-2,098.
105. World Meteorological Organization. 1981. *Measurement of river sediments* Operational Hydraulics Report No. 16. Geneva, Switzerland.
106. Zaslavsky, M. N. 1979. *Erozia Pochv (soil erosion).* Mysl. Publishing House, Moscow, Soviet Union.

4

L. D. Meyer

Rainfall simulators for soil conservation research

Rainfall simulators are research tools designed to apply water in a form similar to natural rainstorms. They are useful for many types of soil erosion and hydrologic experiments. However, rainstorm characteristics must be simulated properly, data analyzed carefully, and results interpreted judiciously to obtain reliable information on the conditions to which the simulated rainstorms are applied.

The major advantages of rainfall simulator research are fourfold: it is more rapid, more efficient, more controlled, and more adaptable than natural rainfall research. Researchers can measure the soil conservation or hydrologic characteristics of newly developed cropping and management practices in a relatively short time. Simulated storms can be applied for selected durations on selected treatment conditions, and measurements from a few such storms often can indicate conclusively at least relative information about those treatments. Plot preparation prior to application of such storms usually takes much less time than plot maintenance for studies depending upon natural rainfall. Frequently, parts of farm fields that have the desired soil characteristics and cropping management histories suffice as research areas. Plots and equipment can be inspected immediately prior to and during data collection. Researchers can make measurements and observations during simulated storms that are difficult or impossible during natural rainstorms. In addition to their use for field studies, rainfall simulators are readily adaptable for highly controlled laboratory research on basic infiltration, runoff, and erosion processes.

But rainfall simulator research also has major limitations and disadvantages. The cost and time required to construct a suitable rainfall simulator and related equipment and the personnel needed to operate an effective simulated rainfall research program generally are the greatest hurdles. The difficulty of simulating natural rainfall characteristics and of properly interpreting data obtained from rainfall simulators that fail to fully achieve such characteristics also impedes such research. Other major problems are the relatively small area to which rain can be applied by most rainfall simulators and the compromise in rainfall characteristics that is necessary for large-area rainfall simulators. These and other limitations need to be recognized and resolved before embarking on a simulated rainfall research effort.

Rainfall simulators do not eliminate the need for natural rainfall experiments. For example, studies that rely on natural rainfall are essential for obtaining data on long-term average erosion or infiltration amounts and on the variability and extremes resulting from climatic differences. However, to be of maximum value, research results must be available as soon as possible after their need is recognized. Conclusive results from field-plot studies that rely on natural rainfall often require many years to obtain a representative sampling of rainstorms that include typical combinations of critical storms and critical treatment conditions. Management practices of interest often change appreciably or may be abandoned during this period. Furthermore, many needed studies never would be initiated if they had to rely on natural rainfall because the results would not be available soon enough. In such situations, rainfall simulators may be not only a useful but an essential component in productive research programs.

Characteristics of rainfall simulators

The ideal rainfall simulator would be inexpensive to build and operate, would simulate rainfall perfectly, would be simple to move, and could be used whenever and wherever needed. Most researchers realize that such a utopian rainfall simulator is impossible to acquire. Thus, different rainfall simulators have different characteristics to meet different research goals. Perhaps the most important characteristics of natural rainfall that need to be closely simulated for soil and water management research are raindrop size distribution, raindrop impact velocity, and appropriate rainstorm intensities. These three characteristics are key factors in soil detachment, soil surface sealing, and resulting runoff. They and other desirable characteristics for rainfall simulators to be used in erosion and hydrologic studies include the following:

▶ *Drop size distribution near that of natural rainstorms.* Natural rainfall consists of a wide distribution of drop sizes that range from near zero to about 7 mm in diameter. The median drop diameter, by volume, is between 1 and 3 mm for erosive rainstorms. Drop diameter generally increases with rain intensity (7).

▶ *Drop impact velocities near those of natural raindrops.* Raindrop fall velocities vary from near zero for mist-sized drops to more than 9 m/sec for the largest sizes. A common-sized raindrop of 2 mm falls at a velocity of 6 to 7 m/sec (4).

▶ *Intensities in the range of storms for which results are of interest.* Intensities of natural rainfall vary from near zero to several millimeters per minute. Generally, very low intensities are not of major interest for erosion and hydrologic studies, and very high intensities are so rare that they may be of limited interest. Intensities between 0.2 and 2 mm/min occur quite commonly and thus are usually of greatest importance.

▶ *Research area of sufficient size to represent satisfactorily the treatments and conditions to be evaluated.* Rainfall simulators should be capable of applying rainfall to plots that are large enough for a realistic test of treatment characteristics. Small plots may be sufficient for studying raindrop impact (interrill) erosion, but much longer plots are necessary for evaluating transport and scour by runoff.

▶ *Drop characteristics and intensity of application fairly uniform over the study area.*

▶ *Raindrop application nearly continuous throughout the study area.*

▶ *Angle of impact not greatly different from vertical for most drops.*

▶ *Capability to apply the same simulated rainstorm again and again.*

▶ *Satisfactory rainstorm characteristics when used during common field conditions, such as high temperatures and moderate winds.*

▶ *Portability for movement from one research site to another.*

Developers of rainfall simulators should strive for these characteristics, especially those that are of greatest importance for their specific uses. But they should not expect to achieve all of the characteristics perfectly. Certain compromises are necessary when simulating rainstorms that are scientifically acceptable in comparison with natural rainstorms. Furthermore, researchers should avoid becoming so involved in developing and improving simulators that little time is left for their use. The goal of rainfall simulator research should be the collection of accurate, useful data, not a perfect rainfall simulator.

Types of rainfall simulators

During the last 50 years, researchers have used a broad range of techniques and equipment for simulating rainfall. These techniques and equipment have ranged from walking up and down the slope with common sprinkling cans to elaborate, pushbutton-operated electronic and hydraulic machines (2, 5, 8, 16, 18). The drop formers that produce the simulated raindrops are the key components of rainfall simulators. Once the drop-producing method is chosen, suitable mechanical and electronic components can be developed for operating them.

The major methods used to produce simulated raindrops for erosion and hydrologic research can be grouped into two rather broad categories: those involving nozzles from which water is forced at a significant velocity by pressure and those where drops form and fall from a tip starting at essentially zero velocity (2, 16, 18). Nozzles produce a wide range of drop sizes, as do rainstorms. But the large nozzle orifices that are necessary to obtain large drops require that the nozzle sprays only intermittently to reduce application rates to typical rain intensities. Tips produce only one drop size or a very limited range of sizes, so they are used mostly for fundamental studies when a carefully controlled drop size is important.

Early rainfall simulators were rather crude compared with today's standards because little information was available on rainfall characteristics. In particular, scientists had not recognized the importance of raindrop impact on soil detachment. Thus, the primary concern was to apply water uniformly over the research area in some manner. Subsequent research evaluating the characteristics of rainstorms provided an understanding of appropriate goals for the design of rainfall simulators. Most of the equipment developed in recent decades has carefully considered raindrop size and velocity data as a basis for more realistic simulation of rainstorms.

Many of the rainfall simulators currently used for soil conservation research are described in the proceedings of the 1979 rainfall simulator workshop (18). A summary of the drop sizes, rain intensities, plot sizes, and literature references for each of them is given on pages 122 to 130 of that report. Some rainfall simulators can apply rainstorms to runoff-plot-sized areas. Others are suitable only for very small field plots or laboratory studies. Figures 1 and 2 show a few of the more widely used designs.

Several rainfall simulators have been designed for use on field plots that are similar in size to those used for natural rainfall studies of runoff and erosion. The rainulator (10) was the first simulator designed to apply rainstorms with drop characteristics near those of natural rainfall on several runoff plots simultaneously. Simulated rain is applied by downward-spraying nozzles that are moved laterally across the plots and border areas. Spray application is intermittent, and only a few intensities can be simulated. The equipment is cumbersome to move from site to site.

The rotating-boom rainfall simulator (17) uses the same nozzles as the rainulator, but the nozzles are located along arms extending from a central vertical shaft that rotates slowly. Only two intensities are possible, and application is intermittent. The entire rainfall simulator is mounted on a trailer and can be readily moved from one site to another.

The programmable rainfall simulator (3) and Kentucky rainfall simulator (13) use rapidly oscillating nozzles that reduce intermittency to short periods. They can produce a wide range of intensities. Both are electronically and mechanically complex. The Kentucky machine is on wheels for more rapid movement to adjacent plots.

Sprinkler irrigation equipment and the Colorado State University RREF rainfall simulator (6) can be used on even larger runoff plots and small watersheds. These types of simulators are less successful in achieving natural rainfall characteristics, especially drop size distribution.

Other rainfall simulators have been designed primarily for research on small field plots of about 1 m² and for laboratory studies. The Purdue sprinkling infiltrometer (1) and rotating-disk rainfall simulator (14) use stationary nozzles, but the latter gives a much better simulation of raindrop-impact energy. The interrill rainfall simulator (11) uses a rapidly oscillating nozzle to produce a wide range of intensities at energies very near natural rainfall. This design also has been adapted for use on longer plots by using additional nozzles (12).

Other designs use yarn or capillary tubing of various materials to form drops, mostly for laboratory studies. Drop sizes are moderate

Figure 1. Examples of rainfall simulators used on large runoff plots: above, rainulator (top) and programmable simulator. Facing page: Kentucky simulator (top) and rotating-boom simulator.

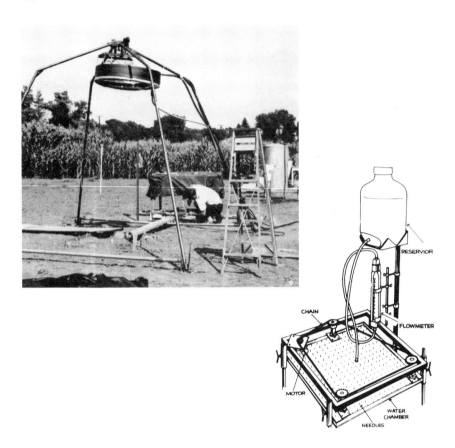

Figure 2. Examples of rainfall simulators used on small field plots or in laboratory studies: Above, rotating-disk simulator (left, top), modular-type infiltrometer (right, center), and variable-intensity laboratory simulator (left, bottom). Facing page: rill/row simulator (top) and interrill simulator.

to large, and all are about the same diameter for any specific design. Impact velocity is considerably less than terminal for heights less than 5 to 10 m. Designs that are being used for erosion research include those by Blackburn, Bubenzer, Gifford, and Romkens (18).

Further details of these and other rainfall simulators can be obtained from referenced publications and from the researchers who developed or now use such equipment. Many of the original designs have been modified to improve the performance or adapt the equipment for other uses.

Any researcher who is considering the development of a rainfall simulator and/or use of simulated rainfall for erosion or hydrologic research should carefully consider the following check-list suggested by Bubenzer (2):

► Define clearly the research objectives.

► Determine if the research objectives can be met reasonably using simulated rainfall.

► Analyze the hydrologic process under study to determine those components that will require accurate simulation.

► Consider plot and soil conditions that will interact with the hydrologic process.

► Determine the short- and long-term economic and labor resources available.

If these considerations suggest that rainfall simulator research is feasible, the researcher should thoroughly study the available literature to see if one or more of the existing rainfall simulators might be suitable. Researchers who are currently using potentially suitable equipment should be contacted to obtain their opinions concerning that simulator for the specific use envisioned and other information concerning its use. Research with rainfall simulators involves many problems and pitfalls, and most researchers are glad to help others avoid problems they have encountered.

Once specific types of equipment have been selected for further consideration, every effort should be made to visit research locations that are using such equipment to see it in operation and learn some of the principles and details first-hand. There is no substitute for a personal encounter with the equipment, its operation, and those operating it. The time and money spent for this purpose can reduce frustrations and save many hours of effort. Even when suitable equipment or designs are not available, a researcher could benefit from discussions with researchers who are experienced in simulated rainfall research before embarking on a rainfall simulator development program.

Personnel requirements and related equipment

Much more is involved in conducting productive simulated rainfall research than simply the availability of a rainfall simulator and the expertise to operate it. A researcher would be unwise to construct a rainfall simulator without the companion commitment for various types of supplementary equipment, plus adequate personnel to conduct effective experimental research once the equipment is completed.

Personnel. The number of persons and necessary skills required to conduct simulated rainfall experiments varies, depending upon the size of the rainfall simulator, the number of samples to be taken, and the analyses to be made at the time of the simulated rainstorms. Most major rainfall simulator field programs need a crew of at least three or four persons, both for assembly and disassembly and to conduct research activities during the tests. Most of their time and effort is required for preparing the plots, installing the runoff measuring and sampling equipment, assembling the rainfall simulator, disassembling the equipment, and completing other measurements. The actual application of the simulated rainstorms is only a small percentage of the total time required for operation, especially for a large rainfall simulator. In addition to field work, considerable time and effort are needed to analyze the great quantity of samples and data normally collected from simulator experiments. Frequently, fieldwork over 2 to 4 months produces enough samples and data to keep researchers busy for the remaining 8 to 10 months of the year.

Pumps, hoses, and water supply. Rainfall simulators often require a large quantity of water, depending upon the area covered and the amount and duration of the rainfall to be simulated. For small rainfall simulators, it may be possible to transport enough water to the site. But large rainfall simulators usually require a water source near the research area. The water supply must be of a suitable quality, both chemically and physically. Wells, clear ponds or lakes, or streams that will be clear at the time of the tests are acceptable water supplies for large rainfall simulators. Because the water usually must be pumped from the source to the rainfall simulator, pumps and hoses of sufficient size are necessary to deliver water at the required pressure and quantity.

Runoff equipment. Equipment is needed to measure runoff rates and to sample the runoff for sediment concentration. Flumes and samplers similar to those used for natural rainfall plots (*15*) can be adapted for use with rainfall simulators. However, they must be modified to make them transportable and easily installed from site to site. On very small plots, all runoff and soil loss may be collected, rather than measuring the runoff and sampling a portion of it. Automated samplers usually are unnecessary because personnel are present during the simulated rainstorms.

Plot borders. The sides and top of rainfall simulator plots must be accurately delineated by a barrier or border, just as for plots tested under natural rainfall. Sheets of metal are frequently used for this purpose. The metal should be driven deep enough into the ground to avoid major subsurface water movement and should extend above the ground far enough that water from outside the plot does not get in and vice versa. Galvanized steel sheets with corrugations that parallel the narrow dimension make suitable borders for large plots because they combine strength with reasonable mass. At the lower end of a plot, collectors are needed to collect the runoff for measurement and sampling.

Rain application measurement. The amount and rate of rain application are important for interpreting the resulting data. Rainfall simulators need to be calibrated accurately for the conditions on which they are used, or the applied rain needs to be measured during the simulated rainstorms. For larger plots, suspended troughs running diagonally across the plot are effective in sampling the variability that occurs at different locations. Small aluminum channels leading to covered containers at their lower end often work quite well for this purpose.

Other equipment. Various other types of ancillary equipment may be necessary to conduct simulated rainfall experiments, depending upon the situation. Movement from site to site may require a truck or other means of transportation. A mobile trailer equipped as a laboratory is useful for analyses during field studies. Numerous containers are required to collect runoff samples, and facilities need to be available to analyze the samples. A portable generator may be required to supply electrical power. Wind shields may be necessary for conducting experiments during moderate winds. Clocks or stopwatches are needed for timing samples, and a camera should be available at all times to document plot conditions. These and other items are important components of a successful rainfall simulator research program.

Simulated rainfall research procedures

In conducting simulated rainfall research, the researcher must make many decisions about conditions to be simulated and the procedures to be followed. These vary from situation to situation, depending upon the purpose of the experiment and the equipment used.

Rainfall intensity variation. Rainfall intensities vary widely during most rainstorms, both with time and space. A rainfall simulator that can vary the intensity during simulated rainstorms is desirable, but this is difficult to achieve with many types of equipment. Variation of intensities from drop formers is difficult but often can be achieved by varying the pressure or water flow to the tips. For nozzle-type equipment, changing the area covered per nozzle or varying the time between successive spray applications can vary the intensity. Even if variations in rainfall intensity during simulated rainstorms can be produced, the combination of intensities and durations to use is difficult to select. A researcher should seriously consider if such efforts are justified, if a specific variable-intensity storm can be repeated identically for each treatment, and if the resulting data can be interpreted meaningfully. In many cases, a constant-intensity storm or a series of constant-intensity storms that each have different intensities is satisfactory. Because data from very low-intensity or very high-intensity storms are usually of limited significance, researchers generally simulate several storm intensities in the range that cause significant erosion or hydrologic events.

Sequence of simulated rainstorms. Throughout the hydrologic year, natural rainstorms occur under a wide range of soil moisture and cover conditions. Seldom can simulated rainfall experiments be conducted for all conditions of interest. The researcher must choose the conditions that will best provide the desired information. Rainstorms sometimes occur on relatively dry soil and at other times on wet to very wet soil. A sequence of storms that evaluates erosion first under relatively dry conditions and, thereafter, under wetter soil conditions usually provides informative data. Furthermore, the initial soil moisture at different research sites seldom can be established at the same level, but the wetness of all plots will be similar and therefore comparable after the initial storm.

If a researcher is studying different rainfall intensities, the obvious preference is to apply each intensity to each of the moisture conditions of interest. However, this often is not feasible because it is time-consuming and requires so many plots. An alternative is to study the effect of rainfall intensity using a series of subsequent storms, at different intensities, on the same research plots where studies of different antecedent moistures were conducted.

Agricultural crops and other land use situations often change greatly through the year. Both plant growth and tillage affect the soil condition and cover that exist at different times throughout the year and, consequently, affect the erosion and hydrologic response. Where such changes are significant, the researcher should attempt to make tests at different times of the year so the trend for major cropping conditions can be determined throughout the year.

Length of simulated rainstorms. The duration of a rainfall simulator test is usually less critical than other decisions. If rainfall intensity-frequency-duration data are available for the area, they may be considered in selecting the rainstorm duration. Another important consideration, of course, is the water supply available for the experiments. Most tests should be long enough so that runoff is well established and infiltration is somewhat constant before rainfall is stopped or its intensity is changed. Generally, this means that the rainstorm at the initial driest condition will be the longest, with succeeding storms somewhat shorter. Because timed runoff and erosion samples generally are collected periodically throughout the simulated rainstorms, the researcher can use the data to determine results from shorter storms than those actually applied.

Comparing simulated rainfall with natural rainfall. Since the characteristics of natural rainfall were established several decades ago, researchers have sought a parameter that would indicate how closely simulated rainfall attained the important characteristics of natural rainfall. The most widely used parameter has been the kinetic energy of rain at impact. Basic physics suggests that kinetic energy, or the similar momentum parameter, should be an important parameter. However, the area over which this energy or momentum is dissipated at impact also may be important. For example, eight raindrops 2 mm in diameter equal the mass of one 4-mm drop. But the horizontal cross section of eight 2-mm drops is twice that of a 4-mm drop. The kinetic energy or momentum of the 2-mm drops, although slightly less due to slower terminal velocity, will be dissipated over twice the area. Thus, the erosiveness of 4-mm drops may be much greater than for several 2-mm drops of the same total kinetic energy. Kinetic energy alone probably is not an adequate parameter for comparison. The conclusions in my 1965 report on possible parameters still seem appropriate: "…until some parameter is proved to be adequate for comparison, this analysis suggests (a) that both the drop-size distribution and drop-fall velocity of natural rainfall should be simulated as closely as possible and (b) that an appreciable sacrifice of either for the other is unwise. One of the parameters may be chosen as a guide, but its influence should be secondary to a comparison with actual raindrop characteristics" (9).

Suitable studies with rainfall simulators. Rainfall simulators are well suited for field studies that compare different soil and cropping conditions. The resulting data give relative values rather than providing absolute rates of erosion. To obtain realistic estimates of annual erosion rates, results from long-term studies under natural rainfall need to be available for at least one of the treatments. The ratio of natural to simulated erosion for that treatment may be assumed for the other conditions to estimate the annual rates.

Suitable studies include those designed to evaluate different crop covers, crop residue management, types and methods of tillage, steepness and length of slope, relative erodibility of different soils, type and sequence of crops in rotations, critical periods in the annual crop cycle, and fundamental mechanics of erosion and runoff. However, treatments that have only minor differences are not appropriate for evaluation by either simulated or natural rainfall because considerable uncontrollable experimental variation usually is associated with such research, especially under field conditions. Therefore, studies with simulated rain should be limited primarily to tests of important conditions that may be expected to produce data with appreciable differences.

Studies of the effects of fertility, rotations, or long-term tillage on erosion rates require plots that are maintained for several years prior to testing. Other studies, such as those comparing a single characteristic of different crops, often can be conducted in a single season. Studies of surface residue rates, cover of a single crop, slope differences, and critical periods often can be made on suitable farm fields. The relative erodibilities of different soils can be studied by applying identical simulated storms to similarily prepared soils at widely varying locations, thereby eliminating the problem of climatic differences among locations, as is present in the K factor of the universal soil loss equation (*19*).

Of course, there are limitations to the types of research that can be conducted with rainfall simulators. Studies that require a wide variety of rainfall intensities or impact energies may not be possible with certain rainfall simulators. Experiments cannot be conducted when water in the lines and fittings will freeze or when crops are at a height that seriously distorts the application and energy patterns. Often, the width and length desired for the research areas are greater than the rainfall simulator can accommodate. Plots with contoured rows generally cannot be evaluated suitably with plot-sized rainfall simulators because the plot borders create unnatural dams that affect the normal water flow; yet, plots with up-and-down-slope rows are not directly applicable to most farm conditions. Nevertheless, treatments for many types of studies can be applied on up-and-down-slope rows to determine the differences among treatments. Such results then can be adjusted for the effects of contouring or other water management practices.

Designing rainfall simulator experiments. Rainfall simulator experiments must be carefully designed statistically to ensure that the results provide the most possible information. Generally, all treatments should be tested in duplicate at least, although additional replicates often are desirable. Proper randomization of treatments and appropriate experimental techniques are also important. Practical considerations may make compromises necessary in the number of treatments versus the number of replications, especially for field studies. The researcher should remember that the goal of the research is useful information about the treatments being studied, not an impressive array of statistics.

Suitable sites. Researchers should consider the following factors in selecting suitable field sites for rainfall simulator tests:
► Past land use history of research area.
► Soil type on research area.
► Past erosion on the research plot area.
► Plot size, shape, slope, and uniformity.
► Availability of water supply.
► Suitability of water supply.
► Availability of and access to the research area.
► Availability of equipment for applying treatments.
► Disposal of runoff from the plot area.
The first four items should be determined accurately and recorded for each study site. Other important site information also needs to be documented.

Research plot conditions. Characteristics of the research plot often affect the suitability of the simulated rainfall. Where the soil surface of all treatments is well covered by plants, mulches, or other surface covers, the impact energy of the simulated rainfall may be of minor importance so far as erosion and infiltration are concerned. Where tall crops, such as mature corn, are growing, the height of the drop-former or nozzle and the type of nozzle movement may affect the suitability of the resulting simulated rain. Where the surface of the plot has a very steep slope and the rainfall simulator is designed to apply all simulated rain from the same elevation, the impact energy and intensity distribution may vary excessively from the upper end to the lower end of a long plot.

Rainfall simulators have been used on plots ranging in size from small cans to greater than a hectare. On areas up to several square meters, elaborate rainfall simulation methods, such as capillary-tubing drop-formers, are feasible. But these methods are not feasible on areas of dozens or hundreds of square meters. For such areas, nozzles that each cover a considerable area are better suited.

When raindrop impact significantly affects the erosion or infiltration rate to be studied, the impact characteristics of rainfall must be simulated adequately. Interrill erosion cannot be evaluated properly unless raindrop impact characteristics are similar to those of appropriate rainstorms. Infiltration rates cannot be evaluated properly on soils that seal from raindrop impact unless the simulated rain acts on these soils in the same manner as natural rainfall.

Results from small areas that are suitable for evaluations of inter-rill erosion and localized infiltration cannot be extrapolated directly to larger areas where flowing runoff would cause rill erosion or where other characteristics may affect infiltration rate. For such conditions, rainfall simulators must apply rainfall to large enough areas so that runoff can accumulate or so that land irregularities affecting infiltration can be evaluated properly.

On soils or topographies where gully erosion or subsurface flow may be major considerations, research areas may need to be quite large. For such studies, it may be most important to apply water relatively uniformly over a very large area, even if it means a sacrifice in the drop impact characteristics. Agricultural irrigation equipment then may be the only feasible approach, particularly if good soil-surface cover is present.

Comparative versus quantitatively accurate results. Some research seeks to evaluate erosion or infiltration quantitatively. Other research is concerned primarily with comparisons among several treatments. For quantitatively accurate results, major rainfall characteristics, such as drop size, impact velocity, and intensities, must be nearly identical to those of rainfall. Where comparison of several conditions is the primary goal, however, studies at one or a limited number of intensities and some compromise in drop characteristics may be acceptable. However, researchers should clearly distinguish for themselves and others the quantitative accuracy of the results.

Laboratory research. Rainfall simulators also are useful for many laboratory studies. Some field rainfall simulators are adaptable for laboratory use. Other rainfall simulators have been designed specifically for laboratory use. The researcher must weigh the advantages of conducting research in the laboratory against the disadvantages of simulated, often quite unnatural conditions necessitated by the laboratory environment. Laboratory studies are not affected by weather and may be conducted throughout the year.

Adjustment for minor differences in application intensity. Identical rates of application generally are impossible when using rainfall simulators at different times under different conditions at different locations. Usually, the variation in intensity is only a few percent, but an adjustment of the resulting runoff and soil loss to a common application intensity is desirable.

For studies where high-intensity rainfall is applied, researchers can assume that a small variation between the actual application intensity and the design intensity will have little, if any, effect on the infiltration rate. Therefore, the actual infiltration amount or rate can be subtracted from the design intensity to obtain the adjusted runoff amount or rate.

A researcher needs to know the effect of intensity on erosion to adjust soil loss determinations. Several analyses and studies have suggested that erosion is approximately proportional to the intensity squared. Unless a different relationship is known to exist for the specific situation, the adjusted soil loss may be computed as follows:

$$\text{Adjusted soil loss} = \left(\frac{\text{selected intensity}}{\text{actual intensity}}\right)^2 \times \text{actual soil loss}$$

Interpretation of rainfall simulator data. The primary result from studies using simulated rainfall is a relative comparison of the treatments during an intense, simulated rainstorm or series of rainstorms. Considerable data interpretation is necessary to obtain indications of average annual values, such as used in the USLE (*19*). The researcher must carefully consider how the relationships among soil erosion and the various independent variables change with time. Extrapolation to field conditions can be made only with careful analysis. However, with proper interpretation, results from rainfall simulator storms potentially can provide useful information in a much more convenient manner than by relying totally on natural rainfall.

Rainfall simulators: Asset or burden?

Rainfall simulators can be great assets to soil erosion and hydrologic research if they are designed and used properly and if the resulting data are interpreted judiciously. However, simulated rainfall is not a magic method for satisfying all erosion and hydrologic research needs. Researchers must weigh carefully both the benefits to be derived from the use of rainfall simulators and the problems that the use of rainfall simulators may cause. Often, simulated rainfall is the only feasible means for conducting needed research. Many research studies could never be considered if they could not be conducted using rainstorms applied by rainfall simulators.

Many decisions need to be made and many questions resolved in considering a research program involving rainfall simulators. But the greatest question to be faced is this: Can the conditions of interest be adequately evaluated using simulated rainfall? Unless a researcher can answer that question positively, with considerable assurance of success, a rainfall simulator can be more of a burden than an asset to a research program. On the other hand, where rainfall simulators are applicable, they can be marvelous research tools in hydrologic and erosion research. The researcher has the responsibility to use appropriate research equipment and techniques that will produce reliable data to help answer important research needs.

REFERENCES

1. Bertrand, A. R., and J. F. Parr. 1961. *Design and operation of the Purdue sprinkling infiltrometer.* Research Bulletin No. 723. Purdue University, West Lafayette, Indiana. 16 pp.
2. Bubenzer, G. D. 1980. *An overview of rainfall simulators.* Paper No. 80-2033. American Society of Agricultural Engineers, St. Joseph, Michigan.
3. Foster, G. R., W. H. Neibling, and R. A. Nattermann. 1982. *A programmable rainfall simulator.* Paper No. 82-2570. American Society of Agricultural Engineers, St. Joseph, Michigan.
4. Gunn, Ross, and G. D. Kinzer. 1949. *Terminal velocity of water droplets in stagnant air.* Journal of Meterology 6: 243-248.
5. Hall, M. J. 1970. *A critique of methods of simulating rainfall.* Water Resources Research 6: 1,104-1,114.
6. Holland, M. E. 1969. *Design and testing of a rainfall system.* CER 69-70, MEH 21. Colorado State University Experiment Station, Fort Collins.
7. Laws, J. O., and D. A. Parsons. 1943. *Relation of raindrop size to intensity.* Transactions, American Geophysical Union 24: 452-460.
8. Meyer, L. D. 1958. *An investigation of methods for simulating rainfall on standard runoff plots and a study of the drop size, velocity, and kinetic energy of selected spray nozzles.* Special Report No. 81. Agricultural Research Service, U.S. Department of Agriculture, Washington, D.C. 43 pp.
9. Meyer, L. D. 1965. *Simulation of rainfall for soil erosion research.* Transactions, American Society of Agricultural Engineers 8: 63-65.
10. Meyer, L. D., and D. L. McCune. 1958. *Rainfall simulator for runoff plots.* Agricultural Engineering 39: 644-648.
11. Meyer, L. D., and W. C. Harmon. 1979. *Multiple-intensity rainfall simulator for erosion research on row sideslopes.* Transactions, American Society of Agricultural Engineers 22: 100-103.
12. Meyer, L. D., and W. C. Harmon. 1985. *Sediment losses from cropland furrows of different gradients.* Transactions, American Society of Agricultural Engineers. 28: 448-453.
13. Moore, I. D., M. C. Hirschi, and B. J. Barfield. 1983. *Kentucky rainfall*

simulator. Transactions, American Society of Agricultural Engineers 26: 1,085-1,089.

14. Morin, J., D. Goldberg, and I. Seginer. 1967. *A rainfall simulator with a rotating disk*. Transactions, American Society of Agricultural Engineers 10: 74-77.

15. Mutchler, C. K., C. E. Murphree, and K. C. McGregor. 1988. *Laboratory and field plots for soil erosion studies*. In Rattan Lal [editor] *Soil Erosion Research Methods*. Soil and Water Conservation Society, Ankeny, Iowa. pp. 9-36.

16. Mutchler, C. K., and L. F. Hermsmeier. 1965. *A review of rainfall stimulators*. Transactions, American Society of Agricultural Engineers 8: 63-65.

17. Swanson, N. P. 1965. *Rotating boom rainfall simulator*. Transactions, American Society of Agricultural Engineers 8: 71-72.

18. United States Department of Agriculture. 1979. *Proceedings of the Rainfall Simulator Workshop, Tucson, Arizona*. ARM-W-10. Northern Plains Soil and Water Research Center, Sidney, Montana. 185 pp.

19. Wischmeier, W. H., and D. D. Smith. 1978. *Predicting rainfall erosion losses— A guide to conservation planning*. Agriculture Handbook 537. Agricultural Research Service, U.S. Department of Agriculture, Washington, D.C. 58 pp.

5

G. R. Foster

Modeling soil erosion and sediment yield

Soil erosion is a serious problem in many countries, especially in such developing countries as India and China. Lester Brown, president of Worldwatch Institute, writes:

"Because of the shortsighted way in which one-third to one-half of the world's cropland is being managed, the soils on this land have been converted from a renewable resource to a nonrenewable one. Assuming an average depth of remaining topsoil of 7 inches, or 1,120 tons per acre, and a total of 3.1 billion acres of cropland, there are 3.8 trillion tons of topsoil with which to produce food, feed, and fiber. At the current rate of excessive erosion, this soil resource is being depleted at the rate of 0.7 percent a year—7 percent each decade. In effect, the world is mining much of its cropland, treating it as a depletable resource not unlike oil" (4).

Although Brown's estimate of the worldwide impact of erosion is probably extreme, his statement expresses a major and shared concern for the loss of crop productivity from long-term soil erosion on cropland. Research has documented erosion's impacts on crop yields and soil degradation (46). Reduced crop yields in the United States because of erosion are not obvious from reported crop yields because technological advances have greatly increased yields over losses caused by erosion (46). Therefore, the impact of past erosion on crop productivity is difficult to assess, and its future impact is even more difficult to estimate because of uncertainties in yield increases from new technology.

Soil erosion causes off-site problems as well as on-site soil degradation. Sediment from erosion on fields can cause downstream sedimentation by filling distant reservoirs or nearby road ditches (1). Sedi-

97

ment in runoff can also pollute receiving waters, and sediment can be a carrier of agricultural chemicals, such as pesticides and plant nutrients used on farm fields (44).

Soil erosion, a diffuse process, occurs at widely varying rates over the landscape, over a field, and even along a typical landscape profile within a field. Therefore, direct measurement of soil erosion at many points across a country, region, or local area is impractical. Physically, erosion is difficult to measure, and variability of climate requires that at least 10 years of data be collected under the best of conditions to obtain an accurate measure of average annual erosion. Many more years are required in arid areas, like the western United States, where large, infrequent storms cause most of the erosion. Consequently, researchers commonly use erosion prediction methods to inventory erosion for national assessments of the impact of erosion on crop productivity, off-site sedimentation, and nonpoint-source pollution (45). For example, recent statements that about one-third of U.S. cropland is eroding excessively are based on erosion prediction at almost 200,000 locations on nonfederal land across the country (45). A similar inventory included these locations plus others, bringing locations sampled to more than a million.

Erosion prediction methods also are useful tools in selecting conservation measures for specific fields. Erosion estimates are made on a specific site for comparisons with the soil loss tolerances assigned to the particular soil where the practices are being planned. Soil loss tolerance is the erosion rate above which erosion is considered to be excessive (51). The Soil Conservation Service has assigned soil loss tolerance values to all major U.S. soils. A satisfactory conservation practice for a specific site is one that gives a predicted erosion rate less than the soil loss tolerance for the soil at that site.

Erosion prediction methods are packages of scientific knowledge that effectively transfer technology from the researcher to the user. They are also convenient tools for extrapolating information where specific field situations have not been studied in research.

Features of an erosion prediction method

Like handtools, a variety of erosion prediction methods are available; each is best at performing a particular task. Therefore, no single prediction method meets all needs. Assessment of the impact of erosion on long-term productivity requires an estimate of long-term average annual erosion. A simple method like the universal soil loss equation directly estimates this erosion without considering individual storms (51). However, the extreme variability of erosive storms in arid areas may require an event-based method that estimates erosion by individual storms to obtain accurate estimates of average annual erosion. Some analyses of nonpoint-source pollution and sediment yield also may require erosion estimates for individual storms.

Erosion varies greatly in space. Sediment yield studies and broad, general regional analyses of erosion often require an estimate of an erosion rate averaged over a large area. In contrast, accurate analyses of the impact of erosion on crop productivity in a particular field requires that erosion and resulting crop yield loss be computed at many points over a landscape. This computation over space is required because erosion varies nonlinearly in space and loss of crop yield varies nonlinearly with erosion rate (32). Therefore, substituting an average erosion rate over a field in erosion-yield loss equations will not always accurately estimate loss of crop yield reductions caused by erosion.

Available resources frequently determine the erosion prediction method used for a given analysis. For example, methods requiring large data bases and mainframe computers cannot be used by a conservationist working directly in the field with a farmer. However, previously computed results from these methods can be displayed in charts and graphs in the office and then carried to the field. In contrast, simple methods like the USLE can be solved directly in the field using slide rules and electronic calculators. However, rapid progress in portable microcomputers and remote communications with mainframe computers is providing computer hardware that makes complex erosion prediction methods available for on-site field applications.

Characteristically, erosion prediction methods are extrapolated beyond the range of the data used to derive them. The ability of a method to perform well when extrapolated is an important factor in the selection of a prediction method, especially in developing countries where little baseline data may exist (17).

Types of soil erosion

The major types of soil erosion by water include sheet, rill, concentrated flow, gully, and stream channel erosion. (This chapter is concerned only with erosion by water, although wind erosion is serious at many locations. See Chapter 10 or Lyles and associates (23) for a discussion of wind erosion processes. Sheet erosion, principally caused by raindrop impact, removes soil in a thin, almost imperceptible layer. Average annual soil erosion rates from raindrop impact may be as great as 40 t/ha (24), much more than a typical soil loss tolerance of 10 t/ha. Rill erosion, caused by surface runoff, results in numerous, small eroded channels across a landscape. Rills are defined as eroded channels so small that tillage operations obliterate them each year. Both sheet and rill erosion are widespread over a field and can exceed 200 t/ha in severe cases. These two types of erosion account for the major impact of soil erosion on crop productivity.

The topography of most fields causes surface runoff to collect in a few major natural waterways before leaving the fields. Erosion that occurs in these areas is called concentrated flow erosion, and the impact of this erosion is localized in and near the waterways. Therefore, loss of soil productivity from concentrated flow erosion is not as great as that from sheet and rill erosion. Farm operators usually till across these eroded channels, and the tilled soil is especially susceptible to soil erosion by flow after seedbed preparation (8). In addition, thawing soil seems to be especially susceptible to concentrated flow erosion.

When eroded channels in concentrated flow areas become so large that they cannot be crossed with farm equipment, they are called gullies. Because modern farm practices require large fields for efficient operation of farm equipment, farmers are careful about letting concentrated flow areas become gullies that would divide a field into smaller units. However, an advancing headcut from off-site can move into and through a field, leaving a large gully. Erosion in stream channels usually does not have impacts on fields except when enlarging stream channels cause serious erosion problems along a field boundary.

Most sediment—eroded soil particles—is carried over and from a field by surface runoff. Sediment carried from a watershed is called sediment yield, and it usually amounts to much less than the sediment produced by erosion within the watershed (1). The difference is deposition, which occurs on the toes of concave slopes, in field boundaries, in low gradient channels, on floodplains, and in reservoirs. The ratio of sediment yield to total erosion within a watershed is called the sediment delivery ratio (SDR), which generally decreases with increases in watershed area (*1*).

Predicting erosion

Researchers must consider the major factors of climate, soil, topography, and land use in predicting erosion (*36*). Obviously, erosion occurs from discrete rainfall events. The amount of rainfall and peak intensity of rainfall are the two most important characteristics of a rainstorm that affect its erosivity. Volume and peak rate of runoff are measures of runoff erosivity. In the United States, climatic erosivity is greatest in the Southeast and lowest in the West, with a ratio of about 20 to 1. Worldwide, erosivity in the United States is greater than that in Europe but less than that in the tropics (*3*). The seasonal distribution of erosivity also varies with location (*51*).

Soils vary in their susceptibility to erosion. A highly erodible soil is about 10 times as erodible as a slightly erodible one. Soil texture (sand, silt, and clay composition), organic matter content, structure, and permeability are major factors that affect erodibility (*51*). Iron, aluminum, and sodium contents also can affect soil erodibility (*37*).

Topographic features of long, steep, and convex slopes can cause severe erosion. Slope steepness affects erosion much more than does slope length. The erosion rate near the end of convex-shaped slopes can be much more than that at the end of a uniform slope of the same average steepness. Conversely, concave slopes can reduce average erosion from a landscape profile by causing large amounts of sediment deposition as well as reduced erosion on the upper portion of these slopes.

The combination of climatic erosivity, soil erodibility, and topography represents the potential erodibility of a site. These factors cannot be changed easily, except in special cases, for example, bench terracing of hillslopes in China (*52*). Land use is the major factor that landowners can manipulate to control erosion. Within land use, cover, management, and supporting conservation practices can be modified. For example, cover can be changed from a cultivated row crop to a close-growing meadow, which will significantly reduce erosion. Even within a cultivated row crop, management can be changed from clean tillage to conservation tillage, in which residue from the previous crop is left on the soil surface to control erosion. Mechanical supporting practices, such as contouring, provide additional erosion control. Often, a field requires a system of practices—conservation tillage to control sheet and rill erosion and terraces and waterways to control concentrated flow erosion and sediment yield.

Worldwide, the most widely used prediction equation for average annual sheet and rill erosion is the USLE (51). That equation is A = RKLSCP, where A is the average annual erosion rate from sheet and rill erosion, averaged over the eroding portion of a landscape profile; R is the factor for climatic erosivity; K is the factor for soil erodibility; LS is the factor for topography; and CP is the factor for land use. A similar equation is SLEMSA—the Soil Loss Estimator for Southern Africa (7). The USLE is an empirical equation derived from more than 10,000 plot-years of data collected on natural runoff plots and an estimated equivalent of 2,000 plot-years of data from rainfall simulators. This equation, in use since the 1960s, is an evolution of several preceding empirical equations dating back to the 1940s (28, 42, 53). The current major USLE guideline manual, *Agriculture Handbook 537* (51), was published in 1978. Application of the USLE is illustrated in the addendum at the end of this chapter.

The Agricultural Research Service and several university scientists currently are revising and updating the USLE. One reason for this revision is to incorporate recent research data, especially for conservation tillage and rangelands, into the equation. Another reason is to improve the applicability of the USLE to climatic regions and land use conditions beyond U.S. cropland east of the Rocky Mountains, the source of the original data for the USLE. Inventory needs of SCS require that the USLE also apply to the western United States and to pastureland, rangeland, and forestland. Currently, the USLE's accuracy under these conditions is somewhat less than that for eastern cropland (10, 43). Another reason to update the USLE is to improve its performance for conditions where no research data exist, such as for vegetable crops. The empiricism of the USLE limits its accuracy when extrapolated to conditions different from those used to derive its factor values.

Improvements being made to the USLE in this revision include using the subfactor method to estimate values for its cover-management factor C (50). This method uses mathematical relationships to describe canopy, ground cover, and within-soil effects (22). Height and percentage of canopy cover that intercepts raindrops are variables in the canopy subfactor. Percentage of ground cover and surface roughness are variables used in the ground cover subfactor. The principal variable for the within-soil effect is the amount of biomass in the upper 100 mm of soil. Both biological ground cover and below-ground biomass are reduced over time by a simple decomposition equation that considers type of biomass, climate, and soil (18).

This updating of the USLE will make increased use of fundamental erosion concepts related to sheet and rill erosion to improve the topographic factor (12) and the subfactor relationship for the effectiveness of ground cover. Also, consideration of the increased susceptibility of thawing soils to erosion is especially important for improving the applicability of the USLE in the Palouse region of the United States. Furthermore, seasonal variability of soil erodibility must be considered when the subfactor method is used to estimate values for the USLE cover-management factor (29).

Major improvements are being made in the western United States to better define erosivity where only limited precipitation data are available to compute erosivity (40). Mountainous areas in the western United States cause major changes in erosivity over distances of only a few kilometers. Lack of adequate precipitation data to derive values for climatic erosivity in a region greatly hampers application of the USLE in other countries (17).

Although most USLE applications are for estimating average annual erosion, the equation is modified and used to estimate soil loss for single-storm events. One of the typical modifications is to use a storm factor based upon rainfall storm energy and maximum intensity, volume of runoff, and peak runoff rate (12, 16). Generation of these runoff values requires a hydrologic model or runoff equation not included in standard USLE procedures. Another typical modification is to use values for the USLE cover-management factor C that apply to the conditions at the time of the storm. The standard C value is a weighted average, based upon the seasonal distribution of rainfall erosivity and seasonal cover-management factor values.

Predicting sediment yield

The topography of many farms causes sediment deposition to occur within fields. Thus, sediment yield is much less than the sediment produced by erosion (34). A simple way to estimate sediment yield is to multiply the USLE erosion estimate by a SDR. Though simple, this approach is not accurate for specific fields because researchers have not determined explicitly the effect of topographic features of fields on SDRs (*34*).

Sediment delivery is a runoff transport process, which makes it highly correlated with the volume of runoff and peak runoff rate. Therefore, scientists frequently use a runoff erosivity factor involving these runoff variables for the erosivity factor in the USLE to estimate sediment yield on a storm-by-storm basis (*47*). Like the event-based USLE, a separate procedure from the USLE is needed to compute runoff values. Also, values applicable at the time of each storm are used for the USLE cover-management factor C.

Erosion and sediment yield prediction methods also can be based on the fundamental concept that sediment yield is determined by either the amount of sediment made available by detachment processes or by the transport capacity of the runoff (*2*, *9*, *19*). Prediction methods based upon this concept usually are complex and require use of an electronic calculator, microcomputer, or large mainframe computer for the most complex methods. These methods include equations for detachment by raindrop impact, detachment by flow, transport by flow, and deposition by flow. These procedures, called models, are a collection of powerful mathematical and logic equations. They can provide information at several locations over the landscape for detachment rates; transport rates; and properties of the sediment being eroded, transported, and deposited. CREAMS, a field-scale model for Chemicals, Runoff, and Erosion for Agricultural Management Systems (*16*, *20*), is one model of this type being used in field applications by SCS. A hydrologic component needed to compute runoff values to drive erosion equations is usually integral with the erosion component in the same computer program. Application of the CREAMS model is illustrated in the addendum.

Predicting concentrated flow and gully erosion

Within the last 5 years, scientists and policymakers have recognized concentrated flow as a major sediment source within fields. SCS is now monitoring this type of erosion at several sites across the United States (*14*). Technology is being developed to predict this type of erosion, and the CREAMS model includes an early version of a component for estimating it (*13*). This component is based on the theory that the erosion rate at a point on the wetted perimeter of a channel is proportional to the difference between the flow's shear stress at that point and the soil's critical shear stress. Several factors, including tillage, soil thawing, and soil texture, greatly affect critical shear stress (8, 9).

Another important feature in many concentrated flows is the effect of untilled soil beneath the tilled surface zone. The more dense, compact, untilled soil can act as a nonerodible layer, restricting the depth and reducing the erosion rate in the concentrated flow channel. The layer also causes eroded concentration flow channels to be wide and shallow, while the channels are narrow and incised when a nonerodible layer is not present.

Little technology exists for predicting gully erosion. What is available is simple and not particularly accurate (1). Most estimates of gully erosion are based on field monitoring.

Future erosion prediction methods

Most future erosion prediction methods will emphasize fundamental and hydrologically based concepts (5, 11, 25, 38, 41). This trend is particularly true in the United States, where rapid advances in portable computers permit use of a complex prediction method in the field. Researchers also are attempting to make erosion equations used in developing countries more fundamentally and hydrologically based (27).

Because most hydrologically based methods compute erosion on a storm-by-storm basis, average annual erosion is computed over 20 or more years of storm values. Stochastic procedures can be used to compute climatic inputs having the statistical properties of historic weather (48). Also, these erosion prediction methods will include simple crop growth models to compute cover and management factors.

Development of large, complex erosion prediction methods requiring mainframe computers will continue. However, these models, such as CREAMS, will be more applicable to a broader range of field conditions. Most of these models currently are limited to research applications. Increased availability of portable computer terminals and telephone communications with mainframe computers will increase the accessibility of these models.

Advancements often increase the difficulty of using erosion prediction technology, which can exceed the user's technical competence. Expert systems, a branch of artificial intelligence, will be developed to help users select an appropriate erosion prediction model, assemble input data required by the model, run the model, interpret model output, and make an assessment or management decision.

Evaluating productivity impacts

Erosion and sediment yield predictions are useful to rank the erosion and sediment yield hazards for several sites or different cropping practices at a site. Although such rankings are useful, erosion estimates compared against soil loss tolerance values provide a quantitative measure of erosion's severity. The soil loss tolerance concept has proven useful for assessing soil erosion and selecting conservation practices for a specific field (51). Current soil loss tolerance values used by SCS are based mainly on scientific information existing prior to the 1960s (39). Because of widespread interest in the impact of soil erosion on crop productivity, researchers are reexamining these values. This has encouraged development of new prediction methods that estimate crop yield reductions due to erosion. One of these, EPIC—Erosion Productivity Impact Calculator—is an elaborate, process-oriented model (48). It includes components for climate, runoff, erosion, soil moisture, soil chemistry, crop growth, management, and economics. Another is the PI—Productivity Index—model (33), which uses a set of simple factor equations for soil properties, available water capacity, resistance to root penetration as indicated by bulk density, and pH. Also, economists are developing models that estimate economic losses associated with erosion (6, 35, 49). The U.S. Department of Agriculture is using such productivity and economic models in its 1985 assessment of the impacts of soil erosion in the United States. Also, current soil loss tolerance concepts and values likely will be adjusted, partially based on results obtained from these models. Applications of the EPIC and PI models are illustrated in the addendum.

Prediction issues

Validity issues. Because no prediction method is perfect, researchers must judge a method's validity for each given application. Each method works best for particular applications. Even for an intended application, estimates will not be exact. But are they close enough? Is theory behind the method valid, and are the results themselves valid, or could they be invalid because of improper input? The validity issue can be addressed by considering applicability, utility, and accuracy.

The prediction method must be applicable to the particular situation. For example, the USLE clearly does not estimate erosion caused by surface irrigation. Therefore, this application of the USLE can be rejected directly as invalid. However, the validity of using the USLE to estimate erosion from sprinkler irrigation is not definite and, therefore, the user must judge whether or not the equation can be applied and determine precautions if it is applied.

The prediction method must be usable. For example, a method may be extremely powerful, but if computing facilities are not available or if research has not determined the required parameter values, the method is not usable and, therefore, inappropriate.

The issue of accuracy usually is addressed by statistical confidence intervals about estimates from the prediction method. However, the required accuracy, expressed by narrowness of the confidence band, depends upon the specific problem. Confidence bands have not been established for most erosion prediction methods, nor have accuracy requirements been established for most applications.

The best measure of validity is this: Does the method serve its intended purpose?

Research issues. Researchers must consider several issues that can affect the final results of research as they plan and conduct further research to develop erosion prediction methods. Obviously, scientists must define the requirements for the prediction method being developed. These objectives must be balanced against available research resources of time, money, personnel, and facilities. If urgency is great or resources are limited, then technology transfer from one location to another is a major consideration. Fortunately, much of the USLE (17) and CREAMS (21) can be transferred to other countries where resources are limited and the need is urgent.

In the United States, high labor costs have reduced significantly the use of natural runoff plots to collect erosion data. Instead, rainfall simulators are now used widely (30). The converse is true in some developing countries, where scientists do not have the resources to build and operate rainfall simulators but can operate natural runoff plots (26). Thus, progress for improving erosion prediction methods can be accelerated by increased international cooperation among erosion scientists.

Development of the governing equations is only one part, actually the smaller part, of developing a new, fundamentally based erosion prediction method. The larger effort is the determination of the parameter values needed to represent the variety of conditions to which a field-applicable method must apply. Though distinct in concept, separating fundamental erosion processes while maintaining their interaction under field conditions is difficult. This hampers research needed to determine parameter values for the fundamentally based prediction methods. Ultimately, a choice must be made for any method, empirical or fundamental, considering whether the difficulty and expense of obtaining additional research data results in a significant improvement in the prediction method.

REFERENCES

1. American Society of Civil Engineers. 1975. *Sedimentation engineering.* New York, New York, 745 pp.
2. Beasley, D. B., E. J. Monke, and L. F. Huggins. 1980. *ANSWERS: A model for watershed planning.* Transactions, American Society of Agricultural Engineers. 23(4): 839-944.
3. Bergsma, E. 1981. *Indices of rain erosivity.* International Institute for Aerial Survey and Earth Sciences. Enschede, The Netherlands. pp. 460-484.
4. Brown, L. R. 1984. *The global loss of topsoil.* Journal of Soil and Water Conservation 39(3): 162-165.
5. Croley, T. E. 1982. *Unsteady overland sedimentation.* Journal of Hydrology 56(3/4): 325-346.
6. Eleveld, B., G. V. Johnson, and R. G. Dumsday. 1983. *SOILEC—Simulating the economics of soil conservation.* Journal of Soil and Water Conservation 38(5): 387-389.
7. Elwell, H. A. 1981. *A soil loss estimation techniques for Southern Africa.* In *Soil Conservation: Problems and Prospects.* John Wiley & Sons. Chichester, England. pp. 281-292.
8. Foster, G. R. 1982. *Channel erosion within farm fields.* Preprint 82-007. American Society of Civil Engineers. New York, New York.
9. Foster, G. R. 1982. *Modeling the soil erosion process.* In *Hydrologic Modeling of Small Watersheds.* American Society of Agricultural Engineers. St. Joseph, Michigan. pp. 297-382.
10. Foster, G. R., J. R. Simanton, K. G. Renard, L. J. Lane, and H. B. Osborn. 1981. *Discussion of application of the universal soil loss equation to rangeland on a per-storm basis.* Journal of Range Management 34(2): 161-165.
11. Foster, G. R., and L. D. Meyer. 1975. *Mathematical simulation of upland erosion by fundamental erosion mechanics.* In *Present and Prospective Technology for Predicting Sediment Yields and Sources.* ARS-S40. Science and Education Administration, U.S. Department of Agriculture, Washington, D.C., pp. 190-207.
12. Foster, G. R., L. D. Meyer, and C. A. Onstad. 1977. *A runoff erosivity factor and variable slope length exponents for soil loss estimates.* Transactions, American Society of Agricultural Engineers. 20(40) 683-687.

13. Foster, G. R., L. J. Lane, J. D. Nowlin, J. M. Laflen, and R. A. Young. 1981. *Estimating erosion and sediment yield on field-size areas.* Transactions, American Society of Agricultural Engineers 24(5): 1,253-1,262.
14. Foster, G. R., L. J. Lane, and W. F. Mildner. 1985. *Seasonally ephemeral cropland gully erosion.* In *Proceedings of the ARS-SCS Natural Resources Modeling Workshop.* Agricultural Research Service, U.S. Department of Agriculture, Washington, D. C. pp. 463-467.
15. Foster, G. R., L. J. Lane, and W. G. Knisel. 1980. *Estimating sediment yield from cultivated fields.* Proceedings of Watershed Management Symposium. American Society of Civil Engineers, New York, New York. pp. 151-163.
16. Foster, G. R., R. E. Smith, W. G. Knisel, and T. E. Hakonson. 1983. *Modeling the effectiveness of on-site sediment controls.* Paper No. 83-2092. American Society of Agricultural Engineers, St. Joseph, Michigan.
17. Foster, G. R., W. C. Moldenhauer, and W. H. Wischmeier. 1982. *Transferability of U. S. technology for prediction and control of erosion in the tropics.* In *Soil Erosion and Conservation in the Tropics.* American Society of Agronomy, Madison, Wisconsin. pp. 135-149.
18. Gregory, J. M., T. R. McCarty, F. Ghidey, and E. E. Alberts. 1983. *Residue decay equation for use in evaluating soil conservation systems.* Paper No. MCR 83-101. Ameican Society of Agricultural Engineers, St. Joseph, Michigan.
19. Khanbilvardi, R. M., A. S. Rogowski, and A. C. Miller. 1983. *Modeling upland erosion.* Water Resources Bulletin 19(1): 29-35.
20. Knisel, W. G., and G. R. Foster. 1981. *CREAMS: A system for evaluating best management practices.* In *Economics, Ethics, Ecology: Roots of Productive Conservation.* Soil Conservation Society of America, Ankeny, Iowa. pp. 177-194.
21. Knisel, W. G., and V. Svetlosanov. 1982. *Conclusions.* In *European and United States Case Studies in Application of the CREAMS Model.* International Institute for Applied Systems Analysis, Laxenburg, Austria. pp. 147-148.
22. Laflen, J. M., G. R. Foster, and C. A. Onstad. 1985. *Simulation of individual-storm soil loss for modeling impact of soil erosion on crop productivity.* In *Soil Erosion and Conservation.* Soil Conservation Society of America, Ankeny, Iowa. pp. 285-295.
23. Lyles, L., L. J. Hagan, E. L. Skidmore. 1983. *Soil conservation: Principles of erosion by wind.* In *Dryland Agriculture.* Agronomy Monograph No. 23. American Society of Agronomy, Madison, Wisconsin. pp. 177-188.
24. Meyer, L. D. 1981. *How rain intensity affects interrill erosion.* Transactions, American Society of Agricultural Engineers 24(6): 1,472-1,475.
25. Meyer, L. D., C. V. Alonso, and W. C. Harmon. 1983. *Modeling soil losses from nearly flat fields.* Paper No. 83-2091. American Society of Agricultural Engineers, St. Joseph, Michigan.
26. Moldenhauer, W. C., and G. R. Foster. 1981. *Empirical studies of soil conservation techniques and design procedures.* In *Soil Conservation: Problems and Prospects.* John Wiley & Sons, Chichester, England. pp. 13-29.
27. Morgan, R.P.C. 1984. *A simple model for assessing annual soil erosion on hillslopes.* In *Proceedings of the International Conference on Agriculture and Environment 1984.* Padua University, Padua, Italy. pp. 1-14.
28. Musgrave, G. W. 1947. *The quantitative evaluation of factors in water erosion, a first approximation.* Journal of Soil and Water Conservation 2(3): 133-138.
29. Mutchler, C. K., and C. E. Carter. 1983. *Soil erodibility variation during the*

year. Transactions, American Society of Agricultural Engineers 26(4): 1,102-1,104, 1,108.

30. Neff, E. L. 1979. *Why rainfall simulation?* In *Proceedings of the Rainfall Simulator Workshop.* Agricultural Research Service, U.S. Department of Agriculture, Washington, D.C. pp. 3-7.

31. Perrens, S. J. 1984. *Evaluation of nonlinearity of crop productivity along nonuniform land slopes.* Agricultural Research Service, National Soil Erosion Laboratory, Purdue University, West Lafayette, Indiana.

32. Perrens, S. J., G. R. Foster, and D. B. Beasley. 1985. *Erosion's effect on productivity along nonuniform slopes.* In *Soil Erosion and Crop Productivity.* American Society of Agricultural Engineers, St. Joseph, Michigan. pp. 201-214.

33. Pierce, F. J., W. E. Larson, R. H. Dowdy, and W.A.P. Graham. 1983. *Productivity of soils: Assessing long-term changes due to erosion.* Journal of Soil and Water Conservation 39(1): 39-44.

34. Piest, R. F., L. A. Kramer, and H. G. Heinemann. 1975. *Sediment movement from loessial watersheds.* In *Present and Prospective Technology for Predicting Sediment Yields and Sources.* Agricultural Research Service, U.S. Department of Agriculture, Washington, D.C., pp. 130-141.

35. Raitt, D. D. 1983. *COSTS—Selecting cost-effective soil conservation practices.* Journal of Soil and Water Conservation 38(5): 384-386.

36. Renard, K. G., and G. R. Foster. 1983. *Theory and principles of soil erosion by water and generalized control strategies.* In *Dryland Agriculture.* Agronomy Monograph No. 23. American Society of Agronomy, Madison, Wisconsin. pp. 155-176.

37. Romkens, M.J.M., C. B. Roth, and D. W. Nelson. 1977. *Erodibility of selected clay subsoils in relation to physical and chemical properties.* Soil Science Society of America Journal 41(5): 954-960.

38. Rose, C. W., J. R. Williams, G. C. Sander, and D. A. Barry. 1983. *A mathematical model of soil erosion and deposition processes: I. Theory for a plane element.* Soil Science Society of America Journal 47(5): 991-995.

39. Schertz, D. L. 1983. *The basis for soil loss tolerances.* Journal of Soil and Water Conservation 38(1): 10-14.

40. Simanton, J. R., and K. G. Renard. 1982. *The USLE rainfall factor for southwestern U.S. rangelands.* In *Proceedings of the Workshop on Estimating Erosion and Sediment Yield on Rangelands.* ARM-W-26. Agricultural Research Service, U.S. Department of Agriculture, Washington, D.C., pp. 50-62.

41. Singh, V. P. 1983. *Analytical solutions of kinematic equations for erosion on a plane. II. Rainfall of finite duration.* Advances in Water Resources 6(June): 88-95.

42. Smith, D. C., and D. M. Whitt. 1947. *Estimating soil losses from field areas of claypan soil.* Soil Science Society of America Proceedings 12: 485-490.

43. Trieste, D. J., and G. F. Gifford. 1980. *Application of the universal soil loss equation to rangelands on a per-storm basis.* Journal of Range Management 33(1): 66-70.

44. U.S. Department of Agriculture. 1980. *CREAMS—a field scale model for chemicals, runoff, and erosion for agricultural management systems.* Conservation Research Report No. 26. Science and Education Administration, Washington, D.C. 643 pp.

45. U.S. Department of Agriculture. 1981. *Soil, water, and related resources in the United States: Status, conditions, and trends. 1980 Appraisal, Part I.* Washington, D.C. 328 pp.

46. Department of Agriculture, National Soil Erosion-Soil Productivity Research Planning Committee (J. R. Williams, chairman). 1981. *Soil erosion effects on soil productivity: A research perspective.* Journal of Soil and Water Conservation 36(2): 82-90.
47. Williams, J. R. 1975. *Sediment-yield prediction with universal equation using runoff energy factor.* In *Present and Prospective Technology for Predicting Sediment Yields and Sources.* ARS-S-40. Agricultural Research Service, U.S. Department of Agriculture, Washington, D.C. pp. 244-252.
48. Williams, J.R., C. A. Jones, and P. T. Dyke. 1984. *A modeling approach to determining the relationship between erosion and soil productivity.* Transactions, American Society of Agricultural Engineers. 27(1): 129-144.
49. Williams, J. R., K. G. Renard, and P. T. Dyke. 1983. *EPIC—A new method for assessing erosion's effect on soil productivity.* Journal of Soil and Water Conservation 38(5): 381-383.
50. Wischmeier, W. H. 1975. *Estimating the soil loss equation's cover and management factor for undisturbed areas.* In *Present and Prospective Technology for Predicting Sediment Yields and Sources.* ARS-S-40. Agricultural Research Service, U.S. Department of Agriculture, Washington, D.C. pp. 118-124.
51. Wischmeier, W. H., and D. D. Smith. 1978. *Predicting rainfall erosion losses.* Agriculture Handbook No. 537. Science and Education Administration, U.S. Department of Agriculture, Washington, D.C. 58 pp.
52. Zhengsan, F., Z. Piehua, L. Qiande, L. Baihe, R. Letian, and Z. Hanxiong. 1981. *Terraces in the loess plateau of China.* In *Soil Conservation: Problems and Prospects.* John Wiley & Sons. Chichester, England. pp. 481-513.
53. Zingg, A. W. 1940. *Degree and length of land slope as it affects soil loss in runoff.* Agricultural Engineering 21(2): 59-64.

ADDENDUM
Example Applications of Erosion and Productivity Models

The USLE model

Soil conservationists working with farmers have used the USLE extensively to guide their choice of conservation practices particularly suited to specific fields. Together, the conservationist and farmer inspect a field and chose the field location where erosion appears to be most severe. A representative landscape profile, together with its slope length and steepness, is chosen for the critical area. For this example, assume 72 m for the length and 6 percent for the steepness, which gives a value of 1.03 for the USLE topographic factor (LS).

The next step is to use a soil survey map to identify the soil on the critical area. Once the soil is identified, values for soil erodibility (K) and soil loss tolerance (T) are selected from a table of soil properties previously developed by soil scientists. Assume that K = 0.042 t·h/MJ·mm and T = 10 t/ha·y for this example. A value for the climatic erosivity factor (R) is read from a map previously prepared from rainfall records of 20 years or more. Assume that R = 2,100 MJ·mm/ha·h·y in this example.

A soil conservation practice is considered satisfactory when computed soil loss at the specific site is equal to or less than the soil loss tolerance for the soil at the site. The cover-management (C) and supporting practices (P) factors in the USLE describe the effectiveness of most soil conservation practices. For this example, assume that tillage is predominantly up-and-down-hill, which sets P = 1.

The USLE is

$$A = R K LS C P \qquad [A1]$$

where A is the computed average annual soil loss in t/ha·y. Because the maximum allowable soil loss is T, this value can be substituted for A so that equation A1 can be rearranged to compute the maximum allowable value C_m for the cover-management factor:

$$C_m = T/R K LS P \qquad [A2]$$

In this example, C_m is computed as:

$$C_m = 10/(2,100 \times 0.042 \times 1.03 \times 1.0) \qquad [A3]$$

$$C_m = 0.11 \qquad [A4]$$

Next, a list of soil conservation practices and their C values are scanned for practices having values less than C_m = 0.11 in this example. The farmer then chooses the practice that he or she prefers. The list of practices and their C values were prepared previously by an agronomist considering variations in plant cover, tillage, prior land use, and other cover-management factors in conjunction with the variation of climatic erosivity over a year. Several aids, including plastic slide rules, are available to facilitate field use of the USLE. Details on the USLE and its application were given by Wischmeier and Smith (52).

The CREAMS model

In contrast to the USLE, which can be used directly in the field, computations with CREAMS are made in the office where the user

Table 1. Analysis of several farming practices for a specific site with CREAMS.

Practice	Sediment Yield (t/ha)	Enrichment Ratio (ER) Based on Specific Surface Area	Product SY·ER (t/ha)
1. Continuous corn, mold-board plow, disk, cultivate, unprotected waterway	16	1.8	28
2. Same as (1), except grassed waterway	5.4	2.7	14
3. Same as (1), except chisel plow, no cultivation, and a grassed waterway	2.7	2.3	6
4. Same as (1), except terraces on a 0.2% grade, and a grassed-outlet channel	3.8	2.8	11
5. Same as (1), except impoundment at lower end of unprotected waterway	1.6	4.2	7

has access to a computer. Whereas the USLE is applied to a representative land profile, CREAMS is applied to a representative watershed of about 5 ha within a field. Topographic data are chosen from a contour map or a field survey for a representative overland flow path, which is usually nonuniform and may include a depositional area. Also, topographic data are chosen for the profile along the concentrated flow channel that typically drains the watershed. Parameter values representing the soil's erodibility and runoff potential are selected from soil survey data. An array of values are selected from the CREAMS manuals (44) and assembled for cover-management parameters related to runoff and erosion for cropstages over a rotation cycle. The remaining data needed to run CREAMS are the climatic file, consisting of daily rainfall over the evaluation period, usually 5 years or more, and long-term average monthly values for temperature and solar radiation. Many of these data files can be reused so that less time is needed to use CREAMS in subsequent applications once the initial data files are prepared.

In a given application, several management alternatives are analyzed with CREAMS. Then results, such as those in table 1, are presented to the farmer. The conservationist and the farmer together select the practice that both provides control of erosion, sediment yield, and chemical loss on sediment and meets the farmer's desires. In this example (Table 1), the major concern is loss of chemicals on sediment from the field. Sediment yield is lowest with practice 5, but the concentration of chemicals on the sediment is highest for this practice, as indicated by the high enrichment ratio. The product of sediment yield and enrichment ratio, the last column in table 1, indicates total loss of chemicals. Because this product is lowest for practice 3, this practice best controls chemical yield associated with sediment. Also, this practice better controls sheet and rill erosion than does practice 5. Details of this example were discussed by Foster and associates (15).

The PI model

Soil loss tolerance values currently used in the United States are based on research conducted before about 1955, and these values also represent the collective judgement of researchers, soil scientists, and others (39). Recent concern about the impact of erosion on productivity has identified the need for an improved, quantitative method for estimating soil loss tolerance values. Assessments of the impact of erosion on productivity increasingly emphasize economic losses due to erosion (45). Such analyses require estimates of yield reductions due to erosion.

The PI model is a simple tool that can be used to guide choice of soil loss tolerance values required to maintain productivity and to estimate yield loss caused by erosion. The model provides no information on potential gully erosion or off-site sedimentation impacts that present soil loss tolerance values consider (39). The PI model uses soil survey data to construct functions of productivity index versus soil loss, such as figure 1 (33). The slopes of the lines in figure 1 represent the vulnerability of a soil to productivity loss caused by erosion. Soils like Port Byron having a flat PI versus eroded depth can tolerate higher erosion rates than can soils like Rockton having a steep PI curve (33).

Use of the PI model to estimate the value of tolerable soil loss is based on an allowable crop yield loss over a planning horizon or time period. For example, assume that loss of 0.1 PI unit over 200 years

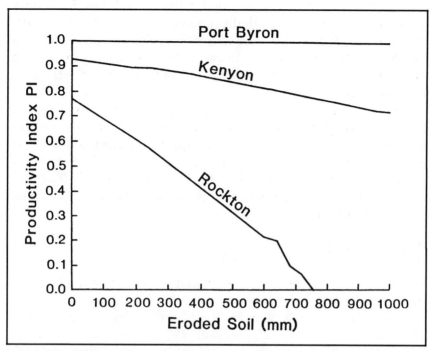

Figure 1. Reduction of the soil productivity index, PI, with soil loss on three soil series in Minnesota; these relationships illustrate the relative vulnerability of soils to erosion.

is acceptable. A very large erosion rate for the Port Byron soil can be tolerated because productivity does not decline with 1 m or more of soil loss. However, other factors, such as off-site sedimentation impact, could require a low soil loss tolerance value for the site in this particular situation. In another situation where off-site sedimentation impact is less critical, the soil loss tolerance value would be higher.

In the case of the Kenyon soil, a loss of 0.1 PI unit represents an eroded soil depth of about 520 mm, which converted to mass per unit area is about 6,760 t/ha, assuming a specific bulk density of 1.3g/cm³ for the soil. This loss, divided by the assumed 200-year planning horizon, gives 35 t/ha·y for the average annual, allowable soil loss. In the case of the Rockton soil, a loss of 0.1 PI unit represents a soil loss of about 110 mm, which converted to mass per unit area is about 1,430 t/ha. This loss divided by the 200-year planning horizon gives 7 t/ha·y for the average annual, allowable soil loss. If the planning horizon is doubled so that productivity loss cannot be greater than 0.1 PI unit over 400 years rather than 200 years, the allowable soil losses will be cut in half. In each situation, factors in addition to productivity loss must be considered in arriving at the final soil loss tolerance value.

When the PI model is used to estimate crop yield loss, an initial PI value is computed for present conditions. The USLE then is used to estimate average annual erosion, which is converted to a loss of soil depth by multiplying by a planning horizon, perhaps 20 years. Another PI value is computed for the reduced soil depth. The ratio of crop PI values equals the ratio of the yield at the end of the period to the current yield.

The EPIC model

EPIC is a powerful, physically based model for computing productivity losses due to erosion at a specific site. It too can be used in the application described for the PI model, including estimates of economic losses caused by erosion (49). A powerful use of EPIC not possible with the PI model is study of the variability in crop yield as a function of climatic variability, which easily masks slight productivity losses caused by erosion.

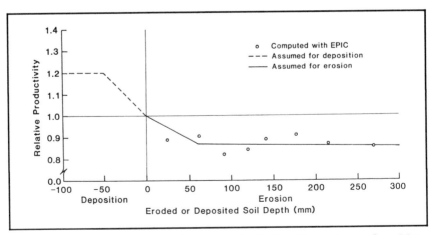

Figure 2. Short-term loss or gain in crop productivity with erosion or deposition, as computed with EPIC.

Table 2. Relative productivity by profile shape and time.

Profile	Relative Productivity by Time* (years)				
Shape	10	20	50	100	200
Uniform	1.00	0.90	0.87	0.86	0.83
Convex	0.88	0.87	0.85	0.83	0.78
Concave	0.99	1.02	1.01	1.01	1.00
Complex	1.00	0.99	0.97	0.95	0.92

*Ratio of net productivity for the profile with erosion and deposition to the net productivity for the profile when no erosion or deposition occurs.

To run EPIC, the user selects parameter values from soil survey data and maps, tables, and information in the EPIC's major components of weather, hydrology, erosion, nutrient, plant growth, soil temperature, tillage, and economics. The model is large and requires significant computer resources. As a consequence, output for a specific soil at a given location is normalized, as shown in figure 2. This figure can be used to analyze the impact of erosion on this soil for any erosion rate without making additional computer runs. Figure 2 from EPIC is similar to figure 1 from the PI model. This procedure was used by Perrens and associates (32) to study the variability of productivity along nonuniform slopes where both erosion and deposition varied greatly. Because EPIC only computes erosion at a single point on the landscape and it does not compute deposition, Perrens and associates (32) combined EPIC and CREAMS, which computes erosion and deposition along nonuniform land profiles. They used EPIC to generate rainfall and runoff values to drive CREAMS. Eroded soil depth computed with CREAMS at a location on the land profile was used in figure 2 to estimate loss of productivity at the location. Values for loss or gain in crop productivity were integrated along the land profile to determine net productivity loss for the profile. Perrens and associates (32) concluded that an accurate estimate of total productivity loss for a field requires consideration of the variation of erosion and productivity over a field, as suggested by table 2.

6

C. W. Rose

Research progress on soil erosion processes and a basis for soil conservation practices

Research objectives in all areas of inquiry develop with time. Early objectives commonly deal with mapping the extent to which the variable of interest (here, the rate or amount of soil erosion per unit area) depends upon the range of factors involved. The methodology developed in the United States that led to the universal soil loss equation (USLE) has met this objective successfully (at least for the U.S. Midwest). But this very success may have delayed development of further objectives in the manner that is common in most areas of research.

Types and objectives of soil erosion models

Scientific investigation of agriculturally related questions often begins with a series of experiments in which each of the variables thought to be important is varied or allowed to vary over a significant range. Then, researchers use statistical models to investigate the data obtained, perhaps leading to a concise summary of the major apparent relationships. This sequence of experimental investigation followed by statistical analysis has occurred in the study of soil erosion carried out by the U.S. Soil Conservation Service. The statistical summary of data from field plot experiments in the U.S. Midwest is the USLE (*11*).

In USLE experiments, the variables of land slope and plot length could be chosen (within limits) and a range of soil types in the geographic region investigated. The experimental program was quite massive because on each soil type the researchers investigated the effect on soil loss of different degrees of cover as well as a suite of land management practices of interest at the time. The significance of environmental and hydrologic characteristics was recognized. Scientists hoped that rainfall characteristics would adequately cover both aspects—a hope not fully fulfilled with the advantages of hindsight.

The USLE summarizes this vast body of regionally derived data, thereby greatly increasing the usefulness of the data base from which it is derived. However, a summary of a data base, whether or not expressed as an equation as in this case, is just that. Thus, researchers increasingly have recognized that the USLE is not universal in its application, partly because it reproduces correlations between rainfall and runoff specific to the data set and partly because of limitations in the range of soil types. Any model based solely on collected data (like the USLE) is a captive of the extent of that data set.

There are more important and general considerations, however. There exists widening recognition that the objective and role of the USLE is not to test a representation of the processes involved in soil erosion. Processes are universal, even though the relative and absolute significance of different processes will vary, as will the particular outcome in any specific set of circumstances. Hence, there is a desire to develop models to represent the processes at work in soil erosion and deposition.

There are quite practical reasons for moving from the purely experimental/statistical approach of the USLE to a process-type approach. First, the USLE deals with "average annual soil loss," a useful concept in the climatic context in which it was developed. For much of the tropical, semitropical, and semiarid world, a far more satisfactory concept is that of a probability distribution of soil loss.

The second reason favoring a move to a process-oriented objective is that in many countries limitations on research resources make it impractical to derive such a probability distribution by direct measurement in all contexts of relevance, despite the historic ability of U.S. scientists to do that for agriculturally important soils in the U.S. Midwest.

Basic approach to sediment erosion and transport

For this discussion, let us restrict consideration to sediment flow on a sloping, planar land surface (Figure 1). Rates of flow per unit

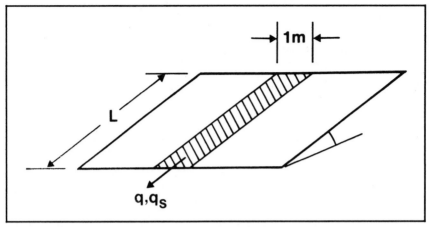

Figure 1. Flow of water (q) and sediment (q_s) from a unit strip width on a planar land element of length L.

strip width of plane are called fluxes (Figure 1). Sediment flux (q_s) is the mass of sediment flowing per unit time across a unit width perpendicular to the direction of the flux. (This sediment mass is expressed on an oven-dry basis.) Likewise, the volumetric water flux (q) is the rate of volume flow of water per unit strip width (in $m^3 \ m^{-1} \ s^{-1}$ or $m^2 \ s^{-1}$).

The sediment concentration (c) is expressed as oven-dry mass of sediment per unit volume of suspension. (All symbols are listed in the addendum.) By definition of these terms, it follows that:

$$q_s = qc \qquad \qquad (kg \ m^{-1} \ s^{-1}) \qquad [1]$$

Soil loss from the land area during an erosion event is calculated by summing the time-variable flux q_s at exit from that area (Figure 1).

From equation 1, it follows that a description of soil erosion processes involves description of the hydrology of surface flow (because of term q) and a description of the various erosion processes that add to sediment concentration, c, and deposition, the only process that tends to reduce c. The magnitude of c arises from the balance between these opposing processes. All of these quantities can vary with time and distance down the plane. However, because average values per unit plane width are used (Figure 1), there is no separate, explicit distinction between rill and interrill processes, even though rilling is a common, although not universal, feature of land surfaces suffering from soil erosion.

The approximate analytic model for overland flow. Rose and associates (*10*) developed the following analytical method for overland flow. Excess rainfall (R) for a land element is defined as follows:

$$R = P - I \qquad\qquad\qquad (m\ s^{-1}) \qquad [2]$$

where P is the rainfall rate and I is the infiltration rate into the land surface (all being functions of time, t).

Let Q represent runoff per unit area. Then, from figure 1:

$$Q = q/L \qquad\qquad\qquad (m\ s^{-1}) \qquad [3]$$

where q is the water flux at x = L, where x is the distance from the top of plane, where it is assumed that q = 0 (Figure 1).

If the land element is small, for example, 1 m², then the excess rainfall is quickly shed by overland flow from the element, in which case:

$$Q = R \qquad\qquad\qquad\qquad (m\ s^{-1}) \qquad [4]$$

However, if the plane length (L) is substantial and R is time variant (as it normally is), then changes in Q will lag behind those in R because of the time taken for water to gather on the soil surface and flow down the plane. Thus, in general, R ≠ Q. Using the approximate analytic theory of Rose and associates (*10*), it may be shown that:

$$R \doteq Q + K_p(dQ/dt) \qquad\qquad (m\ s^{-1}) \qquad [5]$$

where the term K_p depends analytically upon the length, slope, and roughness of the plane, on Q, and on how close the overland flow may be to laminar or turbulent.

For an assumed simple time variation in P, the approximate form of the corresponding relationship between R and Q given by equation 5 is illustrated in figure 2. Note that it follows from equation 5 that R = Q when Q is a maximum (i.e., dQ/dt = 0).

In general, R cannot be measured. However, Q is measured readily, and equation 5 allows R to be calculated from Q. With R known, I can be calculated using equation 2 because P also is measured easily. Hence, infiltration characteristics can be derived allowing I to be estimated from measurements of P (*7*).

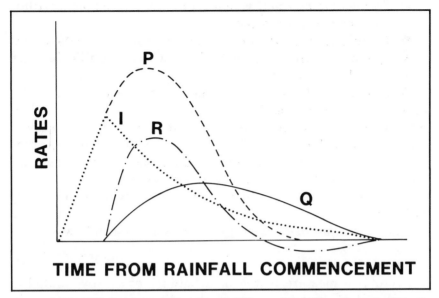

Figure 2. Simplified time-variation in rainfall rate (P), infiltration rate (I), rate of runoff per unit area of plane (Q), and the approximate analytic solution for the excess rainfall rate (R).

The flux q(x) at any x is given from this theory by:

$$q(x) = Qx \qquad\qquad (m^3\ m^{-1}\ s^{-1}) \qquad [6]$$

Erosion/deposition process model

A full description of this model is given by Rose and associates (6, 8, 9). The model relates the sediment flux at any position on a plane and at any time in a runoff event to factors on which this sediment flux depends. The theory also has the capacity, suitably extended, to predict the rate and size distribution characteristics of sediment accumulation elsewhere on the landscape, given information on relevant surface geometry.

Erosion and deposition processes. Excluding landslides or gullies, the following three processes affect sediment concentration:
▶ *Rainfall detachment,* in which raindrops splash sediment from the soil surface into the water of overland flow.

► *Sediment deposition,* which is the result of sediment settling out under the action of gravity.

► *Entrainment of sediment,* the process whereby overland flow picks up sediment from the soil surface, whether in rills, between rills, or in sheet flow without rills.

Detachment and entrainment increase sediment concentration; deposition decreases it, as illustrated in the Forrester-style flow chart of erosion and deposition processes that occur simultaneously at different rates (Figure 3). The resulting sediment concentration (c, Figure 3) is determined by the relative magnitude of these different rates, denoted as e, d, and r, respectively.

The rates of these three processes can be expressed quantitatively as follows:

Rate of rainfall detachment, e. This is given as follows:

$$e = a\ C_e\ P \qquad\qquad (\text{kg m}^{-2}\ \text{s}^{-1}) \qquad [7]$$

where a is a measure of the detachability of soil by rainfall of rate P and C_e is the fraction of the soil surface exposed to the raindrops.

Rate of sediment deposition, d. This rate depends upon sediment size distribution and is very rapid for sand and very slow for clay-sized aggregates or particles. Thus, d must be calculated as the sum of d_i, calculated separately for each sediment size class i with settling velocity v_i. It follows that:

$$d_i = v_i\ c_i \qquad\qquad (\text{kg m}^{-2}\ \text{s}^{-1}) \qquad [8]$$

where c_i is sediment concentration in size class i.

Rate of sediment entrainment, r. The entrainment process in overland flow is similar to bedload transport in streams. The rate of bedload transport can be related to the excess of "stream power" (Ω), above a threshold value (Ω_0) required to entrain sediment (1). Stream power is the rate of working of shear stress between sediment and the streambed. An analogous approach can be developed for r using mass conservation of sediment in the elementary section of overland flow shown in figure 3. The fraction of soil surface, C_r, unprotected from entrainment by overland flow is introduced and plays a similar role to C_e in e (equation 7).

The stream power can be calculated from the bed slope and the water flux q. While the stream power is the maximum rate at which energy is available per unit area, not all this energy is used in entraining and transporting sediment. The efficiency of this conservation is denoted η, where $0 < \eta < 1$.

Model of erosion/deposition on a plane. The model follows from considerations of mass conservation of sediment in the elementary section of overland flow (Figure 3), combined with a marriage of the

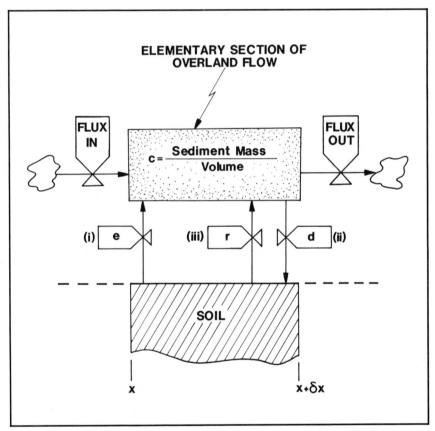

Figure 3. Flow chart (after the style of Forrester) representing the three erosion/deposition processes explicitly represented in the model of Rose and associates (6, 8). Rates of sediment flow are represented by valve symbols: e is rate of rainfall detachment, d is rate of deposition, and r is rate of entrainment of sediment. Fluxes in and out are sediment fluxes entering and leaving the element of flow by overland flow. The elementary section of overland flow is shown artificially elevated above the soil to clarify representation of the sediment fluxes between them. Arrows show the direction of fluxes, and the cloud symbols represent sources and sinks outside the volume of interest.

theories of sediment concentration and hydrology reviewed above.

From figure 3, mass conservation of sediment in size range class i and concentration c_i requires that:

$$\frac{\partial}{\partial x}(q\ c_i) + \frac{\partial}{\partial t}(D\ c_i) = e_i - d_i + r_i \qquad (\text{kg m}^{-2}\ \text{s}^{-1}) \qquad [9]$$

where D is the depth of overland flow at any time and position on the plane and the algebraic sum of rates on the right-hand side of equation 9 represents the net erosion rate.

Using equation 7 (suitably modified) for e_i, equation 8 for d_i, and a more complex expression for r_i, it may be shown that, to a good approximation, the partial differential equation 9 can be reduced to an ordinary (first order) differential equation, which is readily solved. A result of this analysis, analytically summing c_i over all size range classes, yields the sediment concentration [c(L,t)] at the bottom of the plane of length L as a function of time t. The result is:

$$c(L,t) = (aC_eP/QI) \sum_{i=1}^{I} (1/\gamma_i) + \varrho\ gSKC_r(1 - x_\star/L) \quad (\text{kg m}^{-3}),(L > x_\star)[10]$$

The first term on the right-hand side of equation 10 is due to rainfall detachment and the second term to entrainment, both being net values over deposition. The previously undefined terms in this equation include: I, number of sediment size class ranges; γ_i, $1 + v_i/Q$; ϱ, density of water (1,000 kg m^{-3}); g, acceleration due to gravity (9.8 m s^{-2}); S, land slope (sine of inclination angle); K, 0.267η, where η is the efficiency of net sediment entrainment and transport; C_r, the fraction of soil surface unprotected from entrainment by overland flow; and $x_\star$, the distance downslope from the top of the plane beyond which entrainment of sediment commences.

The variable distance $x_\star$ is related to $\mathcal{Q}_0$ (8) by:

$$x_\star = \mathcal{Q}_0/(\varrho\ gSQ) \qquad\qquad (\text{m}) \quad [11]$$

and thus varies with time, as do Q, γ_i, and P.

Soil loss from a plane. From equations 1 and 3, then, at distance downslope x = L, it follows that:

$$q_s(L,t) = c(L,t)QL \qquad\qquad (\text{kg m}^{-1}\ \text{s}^{-1}) \quad [12]$$

The accumulated mass of sediment (M_s) from a plane of width W is thus given as follows:

$$M_s = WL \int_0^{t_R} c(L,t) \, Q \, dt \qquad \text{(kg)} \qquad [13]$$

where t_R is the duration of the runoff event.

In applying equation 13, because sediment concentration $c(L,t)$ and runoff rate Q vary with time, the integral can be adequately approximated by summing over calculations of c repeated at some time interval Δt, which could be the time-averaging period used in some rainfall-rate measuring equipment. Summation may thus require 10 to 20 calculations, which can be carried out by hand calculator, though use of a microcomputer or programmable calculator has obvious advantages.

Methods for obtaining the data required to calculate $c(L,t)$ using equation 10 are given in Rose and associates (6).

If there is a reduction in the slope of the plane, then net deposition will occur. The same theory given above can be modified to yield an expression for the amount, location, and aggregate size distribution of such deposition.

Deposition that occurs in the channel formed by contour banks accumulates with erosion events and can lead to the bank having to be reformed. Deposition of eroded soil in waterways, dams, and other public utilities has a range of economic and social consequences.

A simplified erosion process model

The general model given above can be simplified and still provide a good approximation in many situations.

Equation 10 can be rewritten as follows:

$$c(L,t) = A + B \qquad [14]$$

where A is the net contribution to sediment concentration of rainfall detachment over deposition and B is the net contribution of entrainment over deposition.

The larger the runoff event, that is, the larger Q in equation 10, and the better aggregated the soil, that is, the larger the sedimentary units and so the larger γ_i in equation 10, the smaller is term A compared to B in equation 14. Neglect of term A yields the simplified theory in which sediment concentration is given as follows:

$$c(L,t) = \varrho \, gSKC_r(1 - x_*/L), \qquad\qquad (L > x_*)$$
$$= 2,700 \, S\eta \, C_r(1 - x_*/L) \qquad\qquad (kg \, m^{-3}) \quad [15]$$

because $K = 0.276\eta$.

Concentration $c(L,t)$ in equation 15 is a function of time t only because x_* is time-dependent (through Q, equation 11). If x_* in equation 15 is replaced by a time-averaged mean value, $\bar{x}_*$, defined from equation 11, then:

$$\bar{x}_* = \mathfrak{Q}_0/(\varrho \, gS\bar{Q}) \qquad\qquad\qquad [16]$$

where $\bar{Q}$ is mean rate of runoff per unit plane area, as follows:

$$\bar{Q} = \int_0^{t_R} Q \, dt/t_R$$

The only other term in equation 15 that might be the variable is η, the entrainment efficiency. Assuming this term represents its average value for the erosion event, then sediment concentration can remain constant for a particular erosion event and be given as follows:

$$c \equiv c(L) = 2,700 \, S\eta \, C_r(1 - \bar{x}_* /L) \qquad (kg \, m^{-3}) \quad [17]$$

The terms η and $\bar{x}_*$ (or $\mathfrak{Q}_0$, equation 16) in equation 17 generally are not known and require experimental determination. If length $L > 30$ m very approximately, then $\bar{x}_* /L$ can be small compared to unity, in which case the theory simplifies further to:

$$c = 2,700 \, S\eta \, C_r \qquad\qquad (L > 30 \, m) \, (kg \, m^{-3}) \quad [18]$$

Under rainfall of constant rate, there is experimental support for the constancy of sediment concentration indicated by equations 17 or 18 (3, 4).

Substituting for c from equation 18 into equation 13 yields:

$$\eta = (M_s/WL)/ \, (2,700 \, SC_r \int_0^{t_R} Q \, dt) \qquad\qquad [19]$$

where M_s/WL is total soil loss per unit area and $\int_0^{t_R} Q \, dt$ is total runoff per unit area during the erosion event. If both total losses are measured and $L > 30$ m, then η can be calculated directly from equation 19, provided C_r is also known.

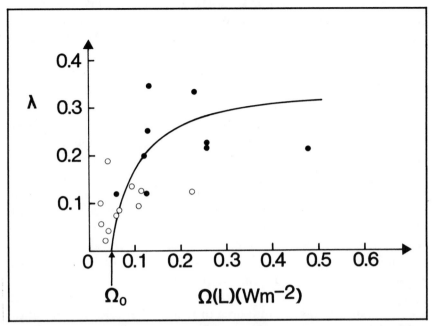

Figure 4. The relationship between the term λ, defined in equation 23, and stream power Ω (L) at exit from the field experimental plots of Dangler and associates (2). Plot slopes varied but were of the same soil type and exposed to simulated rainfall. Plot lengths were either 10.7 m (○) or 22.9 m (●).

If L < 30 m approximately, then $\overline{x}_*$ /L may not be negligible compared to unity, and the form of simplified theory given in equation 17 should be used. More fundamental than a requirement based on slope length L would be a requirement that stream power Ω should be greater than about 0.5 W m^{-2} before equation 18 be used, where:

$$\Omega = \varrho\, gSQL \qquad \qquad \text{(W m}^{-2}\text{)} \quad [20]$$

Justification for the figure of 0.5 W m^{-2} will come later (Figure 4).
Substituting for $\overline{x}_*$ from equation 20, it follows that:

$$c = 2{,}700\ S\ C_r\ \eta\ (1 - \Omega_0/\overline{\Omega}) \qquad\qquad [21]$$
$$= 2{,}700\ S\ C_r\ \lambda \qquad\qquad \text{(kg m}^{-3}\text{)} \quad [22]$$

where

$$\lambda = \eta(1 - \Omega_0/\overline{\Omega}) \qquad\qquad [23]$$

and $\overline{\Omega}$ is a time average value of Ω.

In general, neither η nor Ω_0 is known. Hence, only λ (equation 23) can be calculated from runoff and sediment loss unless $\Omega_0/\overline{\Omega}$ is negligibly small compared to unity. One way in which Ω_0 can be determined is illustrated in the next section.

The effect of plot length on soil loss per unit area

The length (L) of a cultivated plot is an important variable that can be controlled by management. As the scale of mechanical cultivation and harvesting equipment has increased in many countries, so has the length of cultivated slopes between effective barriers to overland flow, such as contour banks. However, especially where cultivation is not mechanized or where the scale of mechanical equipment is modest, then the length of slope between effective barriers to overland flow can be reduced to much smaller values. This is a common practice in Third World countries, and it can lead to substantial reductions in soil loss per unit land area.

The purpose here is to illustrate the application of equation 21 and to relate soil loss per unit area to L and other relevant variables.

Equation 21 can be illustrated using data from Dangler and associates (2), who measured runoff and sediment loss using a rainfall simulator on field soils on the islands of Hawaii and Oahu. Two plot lengths were investigated, 10.7 m and 22.9 m. Simulated rainfall rate was 63.5 mm h⁻¹; experiments lasted about 120 minutes. The first experiment, at prevailing field water content, was sometimes followed some 18 hours later by a second wet run. The data analyzed was for a Molokai soil, a silty clay loam (Typic Torrox, or Oxisol). Prior to these experiments, the test sites had been in continuous sugarcane production.

Figure 4 shows the analysis of this data using equation 22. Despite scatter apparently due to site-to-site variability, a tendency for λ to increase with Ω is evident. Such a relationship would be expected from the form of equation 23, shown fitted as a curve to the data, assuming η is a constant equal to 0.35 and Ω_0 is 0.05 W m⁻². This value (0.05 W m⁻²) corresponds to the value of Ω at which Loch and Donnollan (4) found rilling to commence, accompanied by a quite rapid rise in sediment concentration. It should be noted, however, that their experiments were on quite different soil types than those investigated by Dangler and associates (2). This raises the interesting possibility, requiring further investigation, that Ω_0 may not vary greatly with soil type for soils in a recently cultivated condition.

It follows from equation 23 that λ tends toward the (assumed) constant value of η as Ω increases. Whether or not for any particular bare soil η does have an approximately constant value independent of Ω requires further investigation.

Assuming η is approximately constant, then, despite the scatter in figure 4, the great importance for the factor $(1 - \Omega_0/\Omega)$ in interpreting soil loss from small- to modest-scale experiments is clear.

Using the values of Ω_0 and η obtained by fitting equation 23 to the data in figure 4, we can examine how soil loss would be expected to vary with plot length for a particular suite of variables. From figure 1 and equation 1, the soil loss per unit land area ($m_a = M_s/WL$) is given by $\Sigma q_s/L$. Hence, from equations 13 and 17 and writing $\int_0^{t_R} Q \, dt$ as $\overline{Q}t_R$, then

$$m_a = 2{,}700 \, S \, C_r \, \eta \, (1 - \Omega_0/\overline{\Omega}) \, \overline{Q} \, t_R \qquad \text{(kg m}^{-2}\text{)} \qquad [24]$$

Equation 24 was used to calculate m_a for a range of values of L, the value of $\overline{x}_*$ corresponding to that length at which $\Omega = \Omega_0$. In addition to $\eta = 0.35$ and $\Omega_0 = 0.05$ W m^{-2} from figure 4, a slope of $S = 0.1$ (or 10%) and $C_r = 1$ (bare soil) was assumed. The illustrative values adopted for the hydrologic variables in this calculation correspond to a severe rainstorm:

$$\overline{Q} = 50 \text{ mm h}^{-1} = 1.39 \times 10^{-5} \text{ m s}^{-1}, \text{ and } t_R = 30 \text{ min} = 1{,}800 \text{ s.}$$

The values of m_a calculated from equation 24 using these values are shown plotted against L in figure 5. Notable is the quite rapid rise in m_a with the increase in L beyond $\overline{x}$. (Note that in figure 5 m_a is expressed in t ha^{-1}, where 1 kg m^{-2} = 10 t ha^{-1}). The indication in figure 5 that $m_a = 0$ for $L < \overline{x}_*$ follows from the approximate form of the theory used, which neglects the first term on the right-hand side of equation 10. In practice, this term will ensure some loss, even for $L < \overline{x}_*$. The magnitude of this soil loss for lengths less than that at which entrainment becomes effective requires investigation, but is likely to be typically less than 1 t ha^{-1} for a single rainstorm.

For the particular runoff event and soil characteristics assumed in calculating figure 5, the simple theory predicts that soil loss per unit area would be less than 10 t ha^{-1} only if $L < 6.5$ m approximately. It is the determination of values of η and Ω_0, as illustrated in figure 4, that permits this type of inference to be made. This type of inference can be used in conjunction with information on tolerable rates of soil loss to make recommendations on upper safe limits to plot lengths and how such limits will depend upon land slope, for example.

From figure 5, the use of contour banks or similar structures will reduce soil loss per unit area by reducing the effective value of L. In the context of using large machinery, L may be on the order of 50 m (depending upon slope). In this particular example, the predicted value of m_a is not highly sensitive to values of L in this range. It is not until L is reduced to 7 m that m_a is reduced to about one-half its value at 40 m.

The highest value of η found so far is about 0.7 or 0.8 for cultivated vertisols (Pellusterts and Chromusterts), silt loams (mesic, Typic Fragiudalfs), and loess. Accumulation of the dependence of η (and Ω_0) on soil type and condition is required in the hope that some useful predictive generalizations can be reached. Further experience is also needed on the effect of tillage, tillage type, and time from tillage

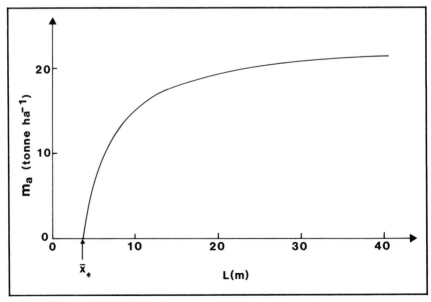

Figure 5. Relationship between soil loss per unit area (m_a) and length of plot (L) for the particular suite of relevant variables given in the text.

on η and Ω_0. The effect of degree of rilling on soil loss also requires more investigation and a better understanding of the processes involved.

Effect of surface contact cover and slope on soil loss

In addition to slope length, land slope (S) and the fraction (C_r) of soil surface not protected by cover in direct contact with it are important factors affecting soil loss. Trade-offs between S and C_r exist in practice at the farm level if the objective of limiting soil loss to some tolerable rate is to be achieved. The tolerable rate of soil loss often is called a T value. If a T value represents a soil loss rate that will not lead to soil deterioration and/or production loss in the long term, it is bound to be quite variable, depending, for example, upon soil depth and all the factors affecting the rate of soil formation. Perhaps partly because of the experimental difficulty in determining T values, there is still argument about the utility of the concept. However, the concept is used here to illustrate the trade-off between the maximum slope that should be cultivated and the level of cover ($1 - C_r$) that can be maintained if the soil loss rate is to be restricted to the T value. It should be noted that only protective material, such as stubble mulch, in contact with the soil surface and protecting it from entrainment is considered as contributing to ($1 - C_r$).

To simplify discussion, assume that $L > 30$ m, approximately. Thus, equation 18 can be used for c, instead of the more general equation 17.

The trade-off is illustrated using the approximate relationship between η and C_r found by Rose and associates (6) and shown in figure 6. Data comes from two different soil types (see legend to figure 6), and there are other causes of scatter. A relationship similar to the type illustrated in figure 6 appears to hold generally.

Accepting a relationship between η and C_r, such as that shown in figure 6, then specifying C_r determines η for the particular soil and cover type from which the data has been obtained. In this context, then, it follows from equation 18 that concentration c can be considered to depend upon only two variables: S and C_r. The dependence of c upon S is direct, but its dependence upon C_r is more complex because of considerable nonlinearity in the relation between η and C_r (Figure 6). The relationship between c and these

two factors for the situation from which figure 6 was derived is shown
in figure 7. It follows from figure 7 or equation 18 with figure 6 that
for given slope and contact cover, c is constant. Thus, from equation 1, the total soil loss per unit width of plane, obtained by summing
over the duration of the runoff event, is given as follows:

$$\int_0^{t_R} q_s \, dt = c \int_0^{t_R} q \, dt \qquad \text{(kg m}^{-1}\text{)} \quad [25]$$

where $\int_0^{t_R} q \, dt$ is the total runoff for the event.

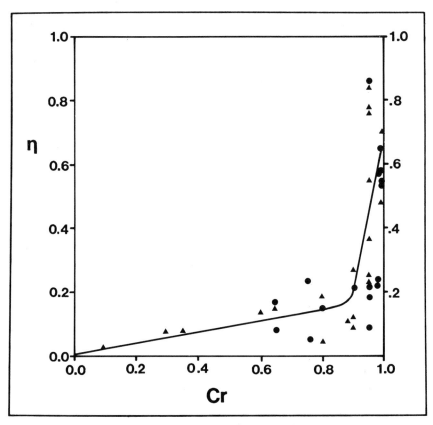

Figure 6. Efficiency of entrainment (η) versus soil surface exposure fraction for two
vertisols in the Darling Downs, Queensland; (▲) refers to a Pellustert, ● to a
Chromustert (6).

For any given site, in general, there is an effect of fractional cover $(1 - c_r)$, not only on c but also on total runoff. Hence, total soil loss is influenced by cover through its influence both on term c and total runoff in equation 25.

For a specific site at Greenmount in the Darling Downs, Queensland, D. M. Freebairn (Queensland Department of Primary

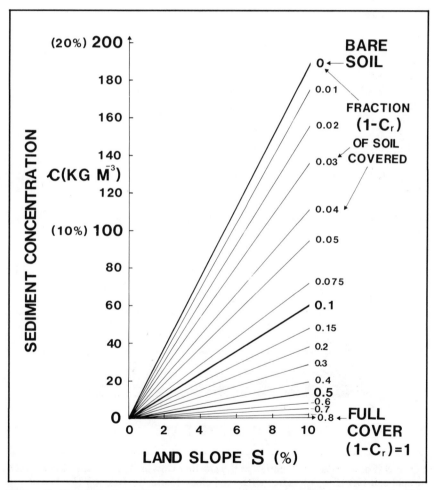

Figure 7. Graph that permits sediment concentrations c (kg m⁻³) to be read off as a function of land slope (S expressed in percent) and fractional soil cover $(1 - C_r)$. Based on the relation shown in figure 6 and approximate equation 18 in the text. Percentages for sediment concentration are approximate.

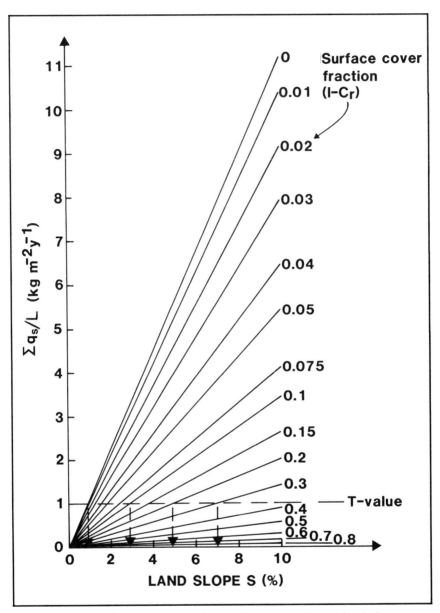

Figure 8. For a particular location in the Darling Downs, Queensland, the average annual soil loss (Σ q_s/L) varies with land slope S and the fraction $(1 - C_r)$ of the soil covered by mulch or other effective contact cover. The T value is a tolerance level corresponding to 1 kg m^{-2} y^{-1} (10 t ha^{-1} y^{-1}). Arrows indicate various trade-offs between maximum cultivated slope and cover if soil loss is not to exceed the T value.

Industries, personal communication) has obtained the following relationship for average annual relationships between runoff and cover:

$$\int_0^{t_R} (q/L)dt = 59 - 29(1 - C_r) \qquad\qquad (\text{mm y}^{-1}) \quad [26]$$

Substituting the suite of values shown in figure 7 for the surface cover fraction $(1 - C_r)$ in equation 26 yields a corresponding suite of values for average annual runoff. Multiplying these by the corresponding value of c (equation 25) gives the average annual soil loss per unit area $(\Sigma\, q_s/L)$ expected at this site for any combination of fractional cover $(1 - C_r)$ and land slope S. Such calculations have been restricted to S = 0.1 (or 10%) because somewhere beyond this limit landslides or other mechanisms involving gravity that are ignored in this theory may become important. The results of these calculations are given in figure 8.

Figure 8 also shows a T value of 1 kg m^{-2} y^{-1} (10 t ha^{-1} y^{-1}), which has been used in some situations. Simply accepting this T value as a desirable upper limit to $\Sigma\, q_s$, then the trade-off can be explored between the maximum land slope that should be cultivated and the fractional cover that can be maintained. Consulting the intersection of the adopted T-value line with the relations in figure 8, we can see that $\Sigma q_s/L < 1$ kg m^{-2} y^{-1} can be achieved for the following illustrative combination of values:

Fractional cover $(1 - C_r)$:	0,	0.1,	0.2,	0.3.
Maximum slope S for cultivation (%):	1,	3,	5,	7.

The fractional cover that can be maintained depends upon many crop and management factors. In mechanized agriculture, cover can be maintained at much higher values if suitable stubble-handling machinery is available that minimizes the burial of stubble from the previous crop. Less intensively mechanized agriculture would appear to be generally compatible with maintaining a high surface cover by stubble. In suitable climates, intercropping, for example, with a shrub legume, also can be an effective soil-conserving practice.

Some conclusions

There are practical reasons for moving from erosion models that summarize a large base of experimental data to models that represent the processes at work in erosion and deposition.

The process model of soil erosion and deposition processes outlined here has received a significant amount and range of testing with field data.

The full model can be substantially simplified and yet maintain adequate accuracy in many situations of significant erosion. Predictive use of this simplified model requires experimental determination of the physically defined parameters η and Ω_0.

The simplified model, especially, appears suitable for use in interpreting experimentation on soil erosion and in the design and assessment of soil-conserving management systems for agriculture at any location, of any cultural type, and for any scale and type of cultivation. The challenge remains to expand experience on values of η and Ω_0 and to seek alternative ways in which these parameters can be measured or predicted.

REFERENCES

1. Bagnold, R. A. 1977. *Bed load transport by natural rivers.* Water Resources 13: 303-311.
2. Dangler, E. W., S. A. El-Swaify, L. R. Ahuja, and A. P. Barnett. 1976. *Erodibility of selected Hawaii soils by rainfall simulation.* Publication ARS-35. Agricultural Research Service, U.S. Department of Agriculture and University of Hawaii, Agricultural Experiment Station, Honolulu.
3. Kilinc, M., and E. V. Richardson. 1973. *Mechanics of soil erosion from overland flow generated by simulated rainfall.* Hydrology Papers No. 63. Colorado State University, Ft. Collins.
4. Loch, R. J., and T. E. Donnollan. 1983. *Field rainfall simulator studies on two clay soils of the Darling Downs, Queensland. I. The effect of plot length and tillage orientation on erosion processes and runoff and erosion rates.* Australian Journal of Soil Research 21: 33-46.
5. Rose, C. W. 1985. *Developments in soil erosion and deposition models.* Advances in Soil Science Vol. 2: 1-63.
6. Rose, C. W., B. R. Roberts, and D. M. Freebairn. 1983. *Soil conservation policy and a model of soil erosion.* In D. E. Byth, M. A. Foale, V. E. Mungomery, and E. S. Wallis [editors] *New Technology in Field Crop Protection.* Australian Institute of Agricultural Science, Melbourne. pp. 212-226.
7. Rose, C. W., D. M. Freebairn, and G. C. Sander. 1984. *GNFIL: A Griffith University program for computing infiltration from field hydrologic data.* School of Australian Environmental Studies Monograph. Griffith University, Brisbane, Queensland.
8. Rose, C. W., J. R. Williams, G. C. Sander, and D. A. Barry. 1983. *A mathematical model of soil erosion and deposition processes: I. Theory for a plane land element.* Soil Science Society of America Journal 47: 991-995.
9. Rose, C. W., J. R. Williams, G. C. Sander, and D. A. Barry. 1983. *A mathematical model of soil erosion and deposition processes. II. Application*

to data from an arid-zone catchment. Soil Science Society of America Journal 47: 996-1,000.

10. Rose, C. W., J. Y. Parlange, G. C. Sander, S. Y. Campbell, and D. A. Barry. 1983. *A kinematic flow approximation to runoff on a plane; an approximate analytic solution.* Journal of Hydrology 62: 363-369.

11. Wischmeier, W. H., and D. D. Smith. 1978. *Predicting rainfall erosion losses— a guide to conservation planning.* Agriculture Handbook No. 537. U.S. Department of Agriculture, Washington, D.C.

ADDENDUM
List of major symbols

Greek/Roman Symbol	Description
a	Detachability of the soil by rainfall
A, B	Terms defined in equation 14
c	Sediment concentration
$c(L)$	Sediment concentration at $x = L$
C_e	Fraction of soil surface unprotected from raindrop detachment
C_r	Fraction of soil surface unprotected from entrainment by overland flow
d	Sediment deposition rate
D	Analytic approximation to depth of overland flow
e	Rainfall detachment rate
g	Acceleration due to gravity
i	As a subscript, refers to a particular sediment size range
I	Number of sediment size ranges, infiltration rate
K	$0.276\ \eta$
K_p	A coefficient depending on the length, slope, and roughness of a plane
L	Length of plane
m_a	Equal to M_s/WL
M_s	Accumulated mass of sediment leaving the plane of width W at $x = L$
P	Rainfall rate
q	Volumetric water flux per unit width of plane
$q(L)$	Value of q at $x = L$, the bottom of the plane
q_s	Sediment flux per unit width of plane
Σq_s	Average annual soil loss per unit area
$q_s(L)$	Value of q_s at $x = L$
Q	Runoff rate per unit plane area
$\bar{Q}$	Time mean value of Q
r	Sediment entrainment rate
R	Excess rainfall rate
S	Slope of the plane (the sine of the angle of land surface inclination)
t	Time
t_R	Duration of runoff event
v_i	Settling velocity of sedimentary particles of size range i
W	Width of plane
x	Distance downslope from the top of the plane
$x\star$	Value of x beyond which $r > 0$
$\bar{x}\star$	Time average value of $x\star$
γ_i	$(1 + v_i/Q)$
η	Efficiency of net entrainment by overland flow $(0 < \eta < 1)$
ϱ	Density of water
Ω	Stream power
Ω_0	Threshold value of Ω

7

R. Lal

Erodibility and erosivity

SOIL erodibility and rainfall erosivity are two important physical factors that affect the magnitude of soil erosion. Erodibility, as a soil characteristic, is a measure of the soil's susceptibility to detachment and transport by the agents of erosion. Erosivity is an expression of the ability of erosive agents to cause soil detachment and its transport. Quantification of these two factors is basic to an understanding of soil erosion processes. The magnitude of soil erosion depends upon the ease with which individual particles are detached by the energy of raindrops and/or overland flow.

Soil erodibility

Soil erodibility is the integrated effect of processes that regulate rainfall acceptance and the resistance of the soil to particle detachment and subsequent transport. These processes are influenced by soil properties, such as particle size distribution, structural stability, organic matter content, nature of clay minerals, and chemical constituents. Soil parameters that affect soil structure, slaking, and water transmission characteristics also affect soil erodibility.

These soil characteristics are dynamic properties. They can be altered over time and under different land uses, soil surface management, and cropping/farming systems. Consequently, soil erodibility also changes over time. Significant alterations in soil erodibility characteristics occur during a rainstorm because of the surface seals or changes in particle orientation that develop.

Soil texture is an important factor that influences erodibility because it affects both detachment and transport processes. While large sand particles resist transport, fine-textured soils resist detachment. The most susceptible textural range for detachment and transport is fine sand and silt. Thus, soils derived from wind-blown parent material, for example, loess, are very susceptible to erosion. The high erosion hazard of loess soils in the catchment of China's Yellow River is a relevant example. Textural and structural properties also influence rainfall acceptance and infiltration capacity. The equilibrium infiltration rate is a function of total porosity, the relative proportion of macropores, and the stability and continuity of macropores. Bio-channels created by decayed roots and soil fauna have major effects on infiltration capacity.

Estimating soil erodibility from laboratory analyses of soil properties. Researchers have used indexes based on routinely measured soil properties to evaluate the relative susceptibility of soils to erosion (Table 1). Most indexes are a measure of a soil's detachability or of its resistance to detachment. Indexes are based on properties that govern aggregation and aggregate stability, water transmission and retention properties, raindrop impact, and thermodynamic processes that govern slaking or disruption of aggregates. These indexes are relative measures of detachability and may not reflect the soil's field behavior in response to rainfall and management. The instability index of De Leenheer and De Boodt (10), the Henin index (20), and the percentage of water-stable aggregates exceeding 0.25 mm are among the indexes related to structural characteristics that have proved applicable in predicting erosion risks for a wide range of soils. These indexes reflect structural properties and a soil's resistance to detachment by rain or overland flow. Also related to the same concept are such indexes as dispersion ratio and colloid ratio. The KE index is a measure of soil strength to resist raindrop impact and is relevant to soil splash.

The choice of an appropriate index depends upon many factors, the most important being the relevance to processes that govern erosion under natural field environments (12). In addition, the suitable index should (a) be simple and easily adapted for routine measurements, (b) be related to other quantifiable soil properties, and (c) be easily used to classify soils into erodibility categories. Considering these factors, there may be no single index that researchers can use for all soils to depict field behavior. A compound factor or combination of several indexes may be required to describe appropriately the field behavior.

Erodibility and the universal soil loss equation. The soil erodibility factor, K, in the USLE is the soil loss from a unit plot per erosion index unit. A unit plot is defined as a 22.1-m length of uniform 9 percent slope, continuously clean-tilled up and down the slope, and maintained absolutely free of vegetative cover (*42*). There are several different methods to measure the erodibility factor:

Natural runoff plots. Preferably, researchers should measure soil erodibility under field conditions. The data base of erodibility measurements using field plots for major benchmark soils is limited and should be strengthened and expanded to regions with severe erosion problems. Methodology and details of runoff and erosion measuring equipment and plot establishment are described in chapter 2.

The surface soil management of the unit plots for measuring erodibility under natural field conditions is based upon the following considerations: (a) plowing up and down the slope to normal depth, followed by disking or harrowing two or three times until a smooth seedbed is achieved; (b) plowing operations are performed at the normal seeding time for the major crops of the region; (c) all other operations involving surface soil disturbance, for example, cultivation to control weeds and eliminate crust formation, should be performed on schedule; and (d) if necessary, herbicides can be used to control weeds.

Plots in warm, tropical climates necessitate more frequent plowing, weed control, and farm operations for seeding and elimination of surface crust than plots in northern latitudes. In tropical regions where bimodal rainfall distribution enables two crops a year, plots also should be plowed twice, once for each growing season. Smoothening of the surface soil also is necessary if rills are progressively deteriorating into gullies. It may be necessary to select a new site after 3 or 4 years if accelerated erosion alters the surface soil and microrelief. Researchers should run these plots for a period of 10 to 20 years because knowledge of long-term trends or alterations in soil erodibility is important for land use planning and for designing conservation measures.

Deviation from unit plot dimensions. It is not uncommon to establish runoff plots for measuring soil erodibility on slopes other than 9 percent and with varying plot sizes to suit local relief char-

Table 1. Soil erodibility indexes based on parameters that can be measured in the laboratory.

Index	Definition	Limits Erodible	Less Erodible	Reference
A. Structural aggregation and stability				
1. Dispersion ratio	amount of (silt + clay) in dispersed state $\times$ 100 / total (silt + clay)	15%	15%	30
2. Silica:sesquioxide ratio	SiO_2/R_2O_3	9.03	0.52	4
3. Clay ratio or mechanical ratio	Sand/(silt + clay)	-	-	5
4. Surface aggregation ratio	Surface area of particles > 0.05 mm / aggregated (silt + clay)	-	-	2
5. Instability index (I)	$I = \Delta$ mean weight diameter in dry and wet seiving	-	-	10
6. Henin index*	$H = (A + L)max/(WS + ES + BS) - 0.9$ S.G	-	-	20
7. Percent water stable aggregates	Percent of water stable aggregates >0.5 mm	-	-	6
B. Water transmission properties				
1. Erosion ratio	dispersion ratio / colloid content/moisture equivalent	>10	<10	30
2. Dispersion-permeability index	$E = KD/AP_p$	-	-	3

*Percent aggregation determined by wet sieving following pretreatment with water (WS), ethanol (ES), and Benzene (BS).

Table 1. Soil erodibility indexes based on parameters that can be measured in the laboratory, continued.

| Index | Definition | Limits | | Reference |
		Erodible	Less Erodible	
C. Water retention properties				
1. Erodibility index (E)	dispersion coefficient × water holding capacity / aggregation	-	-	38
2. Resistance index (I)	soil density × range of particle size / soil moisture content	-	-	8
D. Heat of wetting				
1. Temperature profile during infiltration	ΔT (°C) measured at the wet/dry boundary during infiltration	-	-	9
E. Rain drop technique and rainfall simulator				
1. KE index	Kinetic energy required to disrupt an aggregate at pF 4.44	-	-	7
2. Rainulators	Estimating soil erosion using soil trays subjected to standard rainstorms	-	-	19

acteristics and specific budgetary constraints. Under these circumstances, researchers must adjust soil loss to standard conditions of the unit plot.

The slope factor, LS, is used to correct the factor K for soils of slopes other than 9 percent gradient and 22.1-m length. The LS factor for specific combinations of slope length and gradient may be read directly from the slope-effect chart (Figure 1) or computed by solving the following equation (42):

$$LS = \lambda^{0.5} (0.0076 + 0.0053 \, s + 0.00076 \, s^2) \tag{1}$$

where λ is the field slope length in feet and s is the gradient expressed as slope percent. Similarly, the slope gradient factor (S) and slope length factor (L) can be computed separately by solving the following algebraic equations:

$$S = (0.43 + 0.30 \, s + 0.043 \, s^2)/6.613 \tag{2}$$
$$L = (\lambda/72.6)^{0.5} \tag{3}$$

where s is the gradient expressed as slope percent and λ is slope length in feet. These equations assume uniform slope gradients and do not apply to irregular slopes, such as concave, convex, or complex slope aspects. The irregular slopes are subdivided for computing their LS factor. The dimensions for entering the LS chart or solving the algebraic equations are obtained for the upper segment as such. For the lower segment, however, the steepness of the segment is used with the overall slope length. These equations and charts apply for slope gradients ranging from 2 to 20 percent (42).

Use of rainfall simulation on runoff plots. Establishing and maintaining field runoff plots for a minimum period of 2 to 3 years are capital-intensive and time-consuming operations. Plots can be set up for only a limited number of soils. Soil erodibility data are, however, needed for a vast number of soils of varying physical, chemical, and mineralogical constituents and in diverse relief and climatic environments. It is also important that field measurements of erodibility using runoff plots are related to soil properties. Rainfall simulation can facilitate and expedite data procurement in a relatively short time. Different types of rainfall simulators for field use are described in chapter 4.

Researchers also use rainfall simulators under laboratory conditions to evaluate relative erodibility (11). Antecedent soil moisture content, however, complicates the direct relation between laboratory-

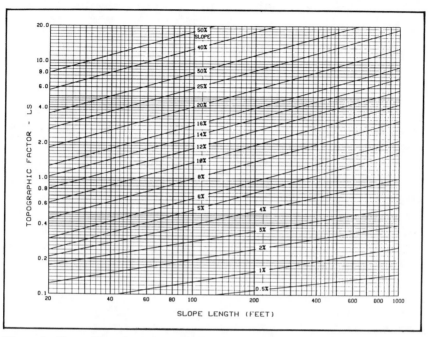

Figure 1. Slope effect chart (topographic factor, LS). LS = (λ/72.6) (65.41 sin$^7\theta$ + 4.56 sinθ + 0.065), where λ = slope length in feet; θ = angle of slope; and m = 0.2 for gradients < 1 percent, 0.3 for 1 to 3 percent slopes, 0.4 for 3.5 to 4.5 percent slopes, and 0.5 for slopes of 5 percent or steeper (*41*).

measured indexes and field behavior. Also important are the magnitude and nature of overland flow and the network of rill systems that develop only under field conditions. In spite of these limitations, a qualitative relationship between laboratory measurements and field responses under natural rainfall conditions can be established (*1*). But a quantitative measure of soil erodibility determined from microplots under laboratory conditions may not be similar to field data because the infiltration characteristics of a shallow layer of soil are not similar to those of a natural, deep soil profile.

Erodibility estimation using a nomogram. Researchers can estimate the erodibility of some soils reliably from soil data if the quantitative relationship between erodibility and soil properties has been established using field runoff plots either under natural or simulated rainfall conditions. Wischmeier and associates (*40*) developed the following algebraic equation relating soil properties and soil erodibility:

$$100K = 2.1 \times 10^4 (12 - OM)M^{1\cdot14} + 3.25(S-2) + 2.5(P-3) \qquad [4]$$

where OM is the percent organic matter, S is the soil structure code (granular, platy, massive, etc.), P is the permeability class, and M is the percent silt and very fine sand. The indexes for soil structure and permeability classes are determined from the *Soil Survey Manual* (36). Figure 2 shows the nomogram solution of equation 4. Soil properties considered in this nomogram are particle size distribution, organic matter content, and qualitative measures of soil structure and water permeability. On the basis of field data available, the silt-size fraction was expanded to include very fine sand. Thus, the particle size classes used in the nomogram are percent silt (0.002-0.05 mm) plus very fine sand (0.05 to 0.10 mm) and percent sand (0.1-2.0 mm).

This nomogram has been tested on the basis of empirical relation-

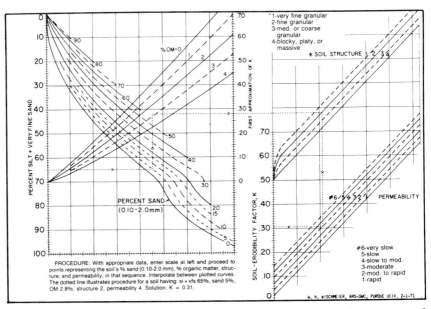

Figure 2. The soil erodibility nomograph. Where the silt fraction does not exceed 70 percent, the equation is 100 K = 2.1 $M^{1.14}$ (10^{-4}) (12 − OM) + 3.25 (S − 2) + 2.5 (P − 3), where M = (percent silt + very fine sand) (100 − percent clay), OM is percent organic matter, S is structure code, and P is profile permeability class (39).

ships between soil characteristics and direct measurements of erodibility for some 13 benchmark soils in the United States. These soils
were mostly of medium texture and of medium to poor structure.
The nomogram also has been tested for heavy-textured soils within
the United States and found suitable by Romkens and associates (34)
and Young and Mutchler (43). The nomogram has not been tested
widely for a broad range of soils outside the region for which it was
developed. El-Swaify and Dangler (14) tested the nomogram for some
Hawaiian soils of volcanic origin and observed that mineralogical
class, which included amorphous constituents, was important in
estimating soil erodibility. Erodibility measurements by Vanelslande
and associates (37) in Nigeria and Ngatunga and associates (32) in
Tanzania indicated that the nomogram requires substantial alterations for application to different soils in the tropics. Soil erodibility
estimated by the nomogram often differs from that measured directly
in the field for Vertisols, Andosols, soils with high gravel contents,
and those with high percentages of iron and aluminum oxides.

Units of erodibility. The soil erodibility factor, K, in the USLE is
the rate of soil loss per unit of R or EI_{30} for a plot 22.1 m long and
of uniform 9 percent slope under continuously clean-tilled fallow.
According to this definition, K has units of mass per area per erosivity
or

$$\frac{mass \cdot area \cdot time}{area \cdot length \cdot force \cdot length} \quad or \quad \frac{(ML^2T)}{(L^2LFL)}$$

In the customary English system, the units are ton $\cdot$ acre $\cdot$ hour/hun-
dreds of acre-foot $\cdot$ ton $\cdot$ inch. In the SI system, the units are metric
ton $\cdot$ hectare $\cdot$ hour/hectare $\cdot$ megajoule $\cdot$ millimeter (t $\cdot$ ha $\cdot$ ha/ha $\cdot$
MJ $\cdot$ mm). The K in SI units is about 0.13 times those of the customary
units.

In summary. Soil erodibility is a dynamic property that is altered
with time due to changes in soil properties. It is a complex parameter
related to many interacting soil characteristics. Erodibility is determined most accurately by direct measurements on field runoff plots
with natural rainfall. The long-term trends involving changes in
erodibility over time should be measured on major soils for 10 to
20 years. Reliable estimates in a shorter time can be obtained by using

rainfall simulators in the field. Indirect estimates of erodibility using predictive relationships are reliable if the basic research information exists regarding the effects of soil properties on erodibility. These statistical relationships should be used for soils similar to those of the original data base. Erodibility indexes based on routinely measured soil properties are of relative importance and provide an indirect measure of a soil's susceptibility to detachment by erosion agents.

Erosivity

The driving force of erosion agents that causes soil detachment and transport is erosivity. The erosivity of rainfall is due partly to direct raindrop impact and partly to the runoff that rainfall generates. The ability of rain to cause soil erosion is attributed to its rate and drop size distribution, both of which affect the energy load of a rainstorm. The erosivity of a rainstorm is attributed to its kinetic energy or momentum, parameters easily related to rainfall rate or total amount.

Momentum. Momentum, a product of mass and velocity, is a measure of the pressure exerted by rainfall on soil. Pressure, or the force per unit area, has the nature of a mechanical stress that causes breakdown or detachment of soil aggregates. That the erosivity of rainfall relates to momentum has been supported by the work of Rose (35) and Williams (39). Statistical relationships between momentum and rainfall parameters developed for Australia by Williams (39) and Kinnell (24) are shown in the following equations:

$$\log \text{ momentum (dynes cm}^2 \text{ h}^{-1}) = 0.711 \log (I) - 1.461 \quad [5]$$
$$\text{momentum (dynes cm}^{-2} \text{ s}^{-1}) = 0.0213 (I) - 0.62 \quad [6]$$

where I is rainfall intensity in mm h^{-1}.

The kinetic energy. The kinetic energy of rainfall is a major factor initiating soil detachment. Direct measurements of kinetic energy can be made with sing pressure transducers or accoustic devices similar to those described by Kowal and Kassam (27). Kinetic energy also can be computed by measuring drop size distribution of rains and assuming terminal velocity corresponding to a given drop size. There are various methods of determining drop size distribution, for example, the flour pellet, stain technique, or oil-capturing methods (22). In addition, many empirical relations have been developed relating kinetic energy to rainfall intensity or rainfall amount. Kinnell (26) described kinetic energy intensity relationships in two ways: (a) the rate of expenditure of the rainfall kinetic energy (E_{RR}), which has the units of energy per unit area per unit time, and (b) the amount of rainfall kinetic energy expended per unit quantity of rain (E_{RA}), which has the units of energy per unit area per unit depth. E_{RA} and E_{RR} are related as follows:

$$E_{RA} = C \ E_{RR} \ I^{-1} \qquad [7]$$

where I is rainfall intensity (depth/time) and C is an empirical constant. Commonly used algebraic equations that relate kinetic energy to rainfall intensity are of the following type:

$$E_{RA} = a + b \log_{10} I \qquad [8]$$
$$E_{RA} = c \ (b - a \ I^{-1}) \qquad [9]$$
$$E_{RA} = bI - a \qquad [10]$$

where I is rainfall intensity and a and b are empirical constants. Some commonly used equations are discussed in the section of this chapter dealing with estimation of rainfall erosivity.

Direct measurement of erosivity. Direct measurements of rainfall erosivity involve monitoring energy load and splash simultaneously. The sand splash is measured for sieved, acid-treated quartz sand of a standard size fraction, maintained at a constant soil moisture potential and packed to a standard density. The standard technique widely used is the Ellison splash cup method (22). The sand splash caused by a rainstorm then is related to simultaneously monitored parameters—kinetic energy, momentum, median drop size, intensity, amount, etc.

Ellison (13) defined standard sand as a fine, round-grained material that passes through a 60-mesh sieve but is retained on a 70-mesh sieve. The sand is treated with hydrogen peroxide to remove organic matter, then is washed clean to remove other colloids and binding materials. The detaching or splashing capacity of a rainstorm is then evaluated using this standard, oven-dried sand packed in small aluminum cups, 8.9 cm (3.5 inches) in diameter and about 5 cm (2 inches) deep (Figure 3). A fine-mesh sieve is fixed at the base with overlying cotton or filter paper to retain the sand but to facilitate water movement. The weight of the oven-dried sand packed in the cup is recorded and then saturated by capillary rise. The cup is then exposed to the rain, and the amount of sand splashed is recorded as the loss in oven-dry weight of sand after it has been splashed by falling raindrops. Numerous modifications and improvements have been made since the original design was published 40 years ago (25).

Erosivity indexes. Attempts have been made to relate detaching or splashing capacity and kinetic energy of rain to routinely measured rainfall parameters, such as rainfall rate and amount, for use in soil loss prediction. Some of the most commonly used indexes include:

Erosivity index (R) of the USLE. Wischmeier and Smith (41) developed a relation between soil loss and a rainfall parameter. The latter is a product (EI_{30}) of the total kinetic energy (E) of the storm

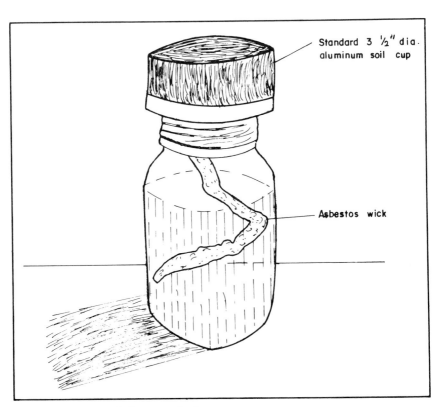

Figure 3. Ellison splash cup technique.

times its maximum 30-minute intensity (I_{30}). This parameter is a compound term that reflects the combined potential of raindrop impact and turbulence created in overland flow. The term I_{30} is computed as twice the greatest amount of rain falling in any 30 consecutive

Table 2. Kinetic energy of rainfall expressed in foot-tons per acre per inch of rain.[*]

Intensity (inch/hour)	0.00	0.01	0.02	0.03	0.04	0.05	0.06	0.07	0.08	0.09
0	—	254	354	412	453	485	512	534	553	570
0.1	585	599	611	623	633	643	653	661	669	677
.2	685	692	698	705	711	717	722	728	733	738
.3	743	748	752	757	761	765	769	773	777	781
.4	784	788	791	795	798	801	804	807	810	814
.5	816	819	822	825	827	830	833	835	838	840
.6	843	845	847	850	852	854	856	858	861	863
.7	865	867	869	871	873	875	877	878	880	882
.8	884	886	887	889	891	893	894	896	898	899
.9	901	902	904	906	907	909	910	912	913	915

	0	0.1	0.2	0.3	0.4	0.5	0.6	0.7	0.8	0.9
1	916	930	942	954	964	974	984	992	1,000	1,008
2	1,016	1,023	1,029	1,036	1,042	1,048	1,053	1,059	1,064	1,069
3	1,074	†								

[*]Computed by the equation, $E = 916 + 331 \log_{10} I$, where E is kinetic energy in foot-tons per acre per inch of rain and I is rainfall intensity in inches per hour.
†The 1,074 value also applies for all intensities greater than 3 inches/hour.

Table 3. Kinetic energy of rainfall expressed in metric ton-meters per hectare per centimeter of rain.[*]

Intensity (cm/h)	.0	0.1	0.2	0.3	0.4	0.5	0.6	0.7	0.8	0.9
0	0	121	148	163	175	184	191	197	202	206
1	210	214	217	220	223	226	228	231	233	235
2	237	239	241	242	244	246	247	249	250	251
3	253	254	255	256	258	259	260	261	262	263
4	264	265	266	267	268	268	269	270	271	272
5	273	273	274	275	275	276	277	278	278	279
6	280	280	281	281	282	283	283	284	284	285
7	286	286	287	287	288	288	289†			

[*]Computed by the equation, $E = 210 + 89 \log_{10} I$, where E is kinetic energy in metric ton-meters per hectare per centimeter of rain and I is rainfall intensity in centimeters per hour.
†The 289 value also applies for all intensities greater than 7.6 cm/h.

minutes. The energy of the storm is calculated by solving the following energy-intensity equation:

$$E = 916 + 331 \log_{10} I \qquad\qquad [11]$$

where E is kinetic energy in foot-tons per acre-inch and I is rainfall intensity in inches per hour. The energy equation of the USLE

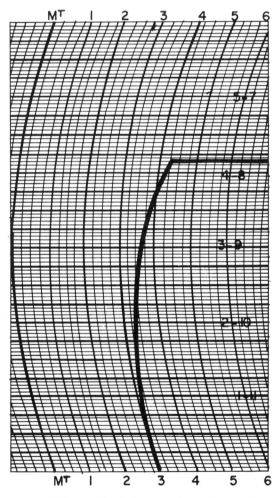

Figure 4. Trace of a daily recording raingage for a
rainstorm on June 16, 1972, at the International Institute for Tropical Agriculture, Ibadan, Nigeria.

expressed in other units is transformed as follows:

$$E = 210.3 + 89 \log_{10} I \qquad [12]$$

where E is in m-t (ha-cm)$^{-1}$ and I is in cm h^{-1}, and

$$E = 11.9 + 8.73 \log_{10} I \qquad [13]$$

where E is in J m^{-2} $-$ mm^{-1} and I is in mm h^{-1}. Tables 2 and 3 provide tabular solutions to the energy equation for a range of intensities.

The kinetic energy of a rainstorm is calculated from the daily recording rain chart (Figure 4) by subdividing the rain into specific intensity ranges. Table 4 illustrates an example computation of EI_{30} index for the rainstorm shown in figure 4. The numerical value of EI_{30} index thus computed varies with the choice of intensity class. The width of intensity class is usually 0.5 inch (10 mm) per hour.

In addition to the energy equation of Wischmeier and Smith (41), other algebraic equations have been developed that relate the energy load of a rainstorm to easily monitored soil parameters, for example, rainfall amount and intensity. Kinnell (24) related kinetic energy to rainfall rate as follows:

$$KE \text{ (ergs cm}^{-2} \text{ s}^{-1}) = 8.37I - 45.9 \qquad [14]$$

where I is the intensity ranging between 0 and 300 mm h^{-1}.

Table 4. Computation of AI_m and EI_{30} index for a rainstorm on June 16, 1972. $E = $ (foot-ton per acre-inch) $= 916 + 331 \log_{10}I$ (where I = inch/hour).

1	2	3	4	5	6	7	8	9	
Time Interval	Rainfall (inch)	Intensity (cm)	Rainfall (in/hr)	Intensity (cm/hr)	AI_m* (cm/hr)	Log I (in/hr)	331 Log I	E	Total E (9 × 2)
7.5	0.75	1.91	6.0	15.24	29.11	0.778	257.5	1,173.5	880.1
7.5	0.65	1.65	5.2	13.21	21.26	0.716	236.9	1,152.9	749.9
7.5	0.55	1.40	4.4	11.18	15.65	0.643	212.8	1,128.8	620.8
7.5	0.45	1.14	3.6	9.14	10.42	0.556	184.0	1,100.0	495.0
7.5	0.25	0.64	2.0	5.08	3.25	0.301	99.6	1,015.6	253.9
7.5	0.55	1.40	4.4	11.18	15.65	0.643	212.8	1,128.8	620.8
7.5	0.15	0.38	1.2	3.05	1.16	0.079	26.1	942.1	141.3
7.5	0.30	0.76	2.4	6.10	4.64	0.380	125.8	1,041.8	312.5
7.5	0.20	0.51	1.6	4.06	2.07	0.204	67.5	983.5	196.1
7.5	0.25	0.64	2.0	5.08	3.25	0.301	99.6	1,015.6	253.9
7.5	0.05	0.13	0.4	1.02	0.13	0.398	197.7	784.3	39.2

*$AI_m = \Sigma \ ai_m = 106.59$; AI_m as obtained from peak intensity and total rainfall amount = 160.93; Total E = 4,563.17; I_{30} (Maximum intensity in 30 minutes) = 4.8 inches/hour; $EI_{30}/100 = (4,563.17 \times 4.8)/100 = 219.03$.

In Zimbabwe, Hudson (21) used the following equation to compute kinetic energy using rainfall intensity I in mm h^{-1}:

$$\text{KE (J m}^{-2}\text{ mm}^{-1}) = 29.8 - \frac{127.5}{I} \qquad [15]$$

In northern Nigeria, Kowal and Kassam (26) related the kinetic energy of a rainfall to the rainfall amount per storm (R_a in mm) as follows:

$$\text{KE (ergs cm}^{-1}) = 41.4\ (R_a - 120.0) \times 10^3 \qquad [16]$$

At Ibadan, Nigeria, I (28) related kinetic energy to rainfall amount P (mm) and to the maximum 30-minute intensity (I_{30} in mm h^{-1}) as follows:

$$\text{KE (J m}^{-2}) = 24.5\ P + 27.6 \qquad [17]$$
$$\text{KE (J m}^{-2}) = 18.2\ I_{30} + 18.2 \qquad [18]$$

In Zimbabwe, Elwell (15) related annual rainfall energy with mean annual rainfall as follows:

$$\text{E (J m}^{-2}\text{ per season)} = 17.368\ (P) \text{ for regions with}$$
$$\text{morning drizzle} \qquad [19]$$
$$\text{E (J m}^{-2}\text{ per season)} = 18.846\ (P) \text{ for regions without}$$
$$\text{morning drizzle} \qquad [20]$$

where P is the mean annual/seasonal rainfall in mm.

The EI_{30} factor has been modified to consider the importance of overland flow. The interaction between the depth of overland flow and the drop diameter causes soil detachment. Soil detachment increases with the increase in depth of overland flow up to a threshold almost equal to raindrop diameter. Therefore, the erosivity term should consist of both rainfall and overland flow or runoff components (31, 33). Onstad and Foster (33) proposed a combined energy term, W, computed as follows:

$$W = 0.5\ EI_{30}\ 15\ Q\ q_p^{1/3} \qquad [21]$$

where Q is storm runoff volume (inches) and q_p is storm peak runoff rate in inches per hour. Further improvements in this equation were made by Foster and associates (17).

Varied and often confusing units of energy are used in the EI_{30} index of the USLE. In customary English units, rainfall rate is measured in inches per hour, kinetic energy in foot-tons per acre-inch or foot-tons per acre, and the storm erosivity EI_{30} index is expressed in units of foot-ton·inch per acre-hour. The corresponding SI units are mm/h for MJ ha·mm^{-1} for energy and MJ·mm ha·hr^{-1} for storm erosivity. The conversion factor for changing storm erosivity from English units to SI units is 0.1702.

KE > 1 index. While working on the rainfall erosivity index for rains in southern Africa, Hudson (22) observed a threshold value of rainfall intensity below which splash was negligible. He observed that erosion occurred only if rain intensity exceeded about 1 inch per hour. He, therefore, developed an index that used kinetic energy of rain segments with intensity exceeding 1 inch per hour (KE > 1). He computed kinetic energy using the intensity-energy equation 15. Hudson (22) obtained better correlation of erosion on soils of Zimbabwe with the KE > 1 index than with EI_{30} index of the USLE. Similar observations have been reported from Sri Lanka by Joshua (23).

AI_m index. Based on soil erosion-rainfall records at Ibadan, Nigeria, I found that for high rainfall intensities in the tropics soil loss was related to the index AI_m minus the product of rainfall amount per storm (A in cm) with the maximum 7.5-minute intensity (I_m in cm h^{-1}) (28). This index is easier to compute than the EI_{30} and KE > 1 indexes and is based on intensity rather than on kinetic energy.

p^2/P index. Fournier (18) developed an erosivity index for river basins on the basis of the relationship between suspended load in rivers and climatic data and relief characteristics. The index, described as climate index C, is defined as follows:

$$C = p^2/P \tag{22}$$

where p is the rainfall amount in the wettest month and P is the annual rainfall amount. This index subsequently was modified by FAO (16) as follows:

$$C_1 = \sum_{i=1}^{12} \frac{p_i^2}{P} \tag{23}$$

where p_i is the rainfall in a month and P is the annual rainfall. This index summed for the whole year was found to be linearly correlated with EI_{30} index (R) of the USLE as follows:

$$R = b + a(C_1) \tag{24}$$

where the constants a and b vary widely among different climatic zones. The value of intercept b is -152, -420, -3, and -416 for the United States, the eastern United States, the western United States, and West Africa, respectively. The corresponding values of coefficient a are 4.17, 6.86, 0.66, and 5.44, respectively. This index is an approximation of the EI_{30} index for regions where long-term recording raingage records are not available.

In summary. Erosivity is best estimated by direct measurements of a rainstorm's energy load. The data base for these measurements is limited to a few regions only and must be expanded to other agriculturally important areas. Empirical equations that relate rainfall energy with intensity are needed urgently, especially for tropical regions characterized by high-intensity rainstorms. The reliability of the various erosivity indexes discussed depends upon the basic data available. The EI_{30} index can be used reliably for a wide range of climatic regions if the intensity-energy equations for the region are available. Adopting any index to estimate erosivity without evaluating its applicability can lead to grossly erroneous estimates of soil loss.

REFERENCES

1. Aina, P. O., R. Lal, and G. S. Taylor. 1979. *Relative susceptibility of some Nigerian soils to water erosion.* Nigerian Journal of Soil Science 1: 1-18.
2. Anderson, H. W. 1954. *Suspended sediment discharge as related to streamflow, topography, soil, and land use.* Transactions, American Geophysical Union 35: 268-281.
3. Baver, L. D. 1933. *Some factors affecting erosion.* Agricultural Engineering 14: 51-52.
4. Bennett, H. H. 1926. *Some comparisons of the properties of humid-tropical and humid-temperate American soils, with special reference to indicated relations between chemical composition and physical properties.* Soil Science 21: 349-375.
5. Bouyoucos, G. J. 1935. *The clay ratio as a criterion of susceptibility of soils to erosion.* Journal of American Society of Agronomy 27: 738-741.
6. Bryan, R. B. 1968. *Development, use and efficiency of indices of soil erodibility.* Geoderma 2: 5-26.
7. Bruce-Okine, E., and R. Lal. 1975. *Soil erodibility as determined by raindrop technique.* Soil Science 119: 149-157.
8. Chorley, R. J. 1959. *The geomorphic significance of some Oxford soils.* American Journal of Science 257: 503-515.
9. Collis-George, N., and R. Lal. 1971. *Infiltration and structural changes as influenced by initial moisture content.* Australian Journal of Soil Research 9: 167-116.
10. De Leenheer, L., and M. De Boodt. 1959. *Determination of aggregate stability by the change in mean weight-diameter.* Mededelingen van de Landbouwhogeschool, Gent 24: 290-351.

11. De Ploey, J., and D. Gabriels. 1980. *Methods of measuring soil loss.* In
    M. J. Kirkby and R.P.C. Morgan [editors] *Soil Erosion.* J. Wiley & Sons,
    Chichester, England.
12. De Vleeschauwer, D., R. Lal, and M. De Boodt. 1978. *Comparison of
    detachability in relation to soil erodibility for some important Nigerian soils.*
    Pedologie 28: 5-20.
13. Ellison, W. D. 1947. *Soil erosion studies. Part II. Soil detachment hazard by
    raindrop splash.* Agricultural Engineering 28: 197201.
14. El-Swaify, S. A., and E. W. Dangler. 1977. *Erodibility of selected tropical
    soils in relation to structural and hydrologic parameters.* In *Soil Erosion: Predic-
    tion and Control.* Soil Conservation Society of America, Ankeny, Iowa. pp.
    105-114.
15. Elwell, H. A. 1978. *Destructive potential of Zimbabwe/Rhodesia rainfall.*
    Rhodesian Agriculture Journal 76: 227-232.
16. Food and Agriculture Organization, United Nations. 1977. *Assessing soil
    degradation.* Soils Bulletin 34. Rome, Italy.
17. Foster, G. R., L. D. Meyer, and C. A. Onstad. 1977. *A runoff erosivity factor
    and variable slope length exponents for soil loss estimates.* Transactions,
    American Society of Agricultural Engineers. 20: 683-687.
18. Fournier, F. 1960. *Climat et erosion: la relation entre l'erosion du sol par
    l'eau et les precipitations atmospheriques.* Presses Universitaies de France, Paris.
19. Gabriels, D. and M. DeBoodt. 1975. *A rainfall simulator for erosion studies
    in the laboratory.* Pedologie 25:80-86.
20. Henin, S., G. Mournier, and A. Combeau. 1958. *Methode pour l' etude de
    la stabilite structurale des sols.* Anales Agronomiques 1: 71-90.
21. Hudson, N. W. 1965. *The influence of rainfall on the mechanics of soil ero-
    sion with particular reference to northern Rhodesia.* M.S. thesis. University
    of Cape Town, Cape Town, South Africa.
22. Hudson, N. W. 1976. *Soil conservation.* BT Batsford, London, England.
23. Joshua, W. D. 1975. *Erosive power of rainfall in the different climatic zones
    of Sri Lanka.* Proceedings, Symposium of Erosion and Solid Matter Transport
    in Inland Waters. Publication 127. International Association of Hydrological
    Sciences, Wallingford, England.
24. Kinnell, P.I.A. 1973. *The problem of assessing the erosive power of rainfall
    from meteorological observations.* Soil Science Society of America Proceedings
    37: 617-621.
25. Kinnell, P.I.A. 1974. *Splash erosion: Some observations on the splash-cup
    technique.* Soil Science Society of America Proceedings 38: 657-660.
26. Kinnell, P.I.A. 1981. *Rainfall intensity-kinetic energy relationships for soil loss
    prediction.* Soil Science Society of America Proceedings 45: 153-155.
27. Kowal, J. M., and A. H. Kassam. 1976. *Energy and instantaneous intensity
    of rainstorms at Samaru, northern Nigeria.* Tropical Agriculture 53: 185-198.
28. Lal, R. 1976. *Soil erosion problems on an Alfisol in western Nigeria and their
    control.* Monograph 1. International Institute of Tropical Agriculture, Ibadan,
    Nigeria. 208 pp.
29. Lal, R. 1981. *Analysis of different processes governing soil erosion by water
    in the tropics.* Publication 133. International Association of Hydrological
    Sciences, Wallingford, England. pp. 351-364.
30. Middleton, H. E. 1930. *Properties of soils which influence soil erosion.* Technical
    Bulletin 178. U.S. Department of Agriculture, Washington, D.C. 16 pp.
31. Monke, E. J., H. J. Marelli, L. D. Meyer, and J. F. De Jong. 1977. *Runoff,*

erosion and nutrient movement from interill areas. Transactions, American Society of Agricultural Engineers 20: 58-61.

32. Ngatunga, E.L.N., R. Lal, and A. P. Uriyo. 1984. *Effects of surface management on runoff and soil erosion from some plots at Mlingano, Tanzania.* Geoderma 33: 1-12.

33. Onstad, C. A., and G. R. Foster. 1975. *Erosion modelling on a watershed.* Transactions, American Society of Agricultural Engineers 18: 288-292.

34. Römkens, M.J.M., D. W. Nelson, and C. B. Roth. 1975. *Soil erosion on selected high clay subsoils.* Journal of Soil and Water Conservation 30: 173-176.

35. Rose, C. W. 1960. *Soil detachment caused by rainfall.* Soil Science 89: 28-35.

36. U.S. Department of Agriculture. 1951. *Soil survey manual.* Agriculture Handbook No. 18. Washington, D.C.

37. Vanelslande, A., P. Rousseau, R. Lal, D. Gabriels, and B. S. Ghuman. 1984. *Testing the applicability of soil erodibility nomogram for some tropical soils.* Publication 144. International Association of Hydrological Sciences, Wallingford, England. pp. 468-473.

38. Vosnesensky, A. S., and A. B. Artsruui. 1940. *A laboratory method for determining the anti-erosion resistance of soils.* Soils and Fertilizer 10: 289.

39. Williams, M. A. 1969. *Prediction of rainfall splash erosion in the seasonally wet tropics.* Nature 222: 763-765.

40. Wischmeier, W. H., C. B. Johnson, and B. V. Cross. 1971. *A soil erodibility nomogram for farmland and construction sites.* Journal of Soil and Water Conservation 26: 189-192.

41. Wischmeier, W. H., and D. D. Smith. 1958. *Rainfall energy and its relationship to soil loss.* Transactions, American Geophysical Union 39: 285-291.

42. Wischmeier, W. H., and D. D. Smith. 1978. *Predicting rainfall erosion losses— a guide to conservation planning.* Agriculture Handbook 537. U.S. Department of Agriculture, Washington, D.C. 58 pp.

43. Young, R. A., and C. K. Mutchler. 1977. *Erodibility of some Minnesota soils.* Journal of Soil and Water Conservation 32: 180-182.

8

M. A. Stocking

Assessing vegetative cover and management effects

$\mathbf{M}$ANY people believe that vegetative cover is the single most important factor in soil erosion control in the tropics. Soil conservation systems increasingly emphasize the role of organic matter, dead or alive, in arresting erosion. Management for soil conservation is now as much a question of encouraging vegetative growth as it is of constructing physical conservation measures.

As a strategy for soil conservation planning, the promotion of vegetation, or a "biological" approach to soil conservation, has much to offer. Vegetation is the factor most easily manipulated by careful management. Beyond that, better vegetative growth and, hence, better protection of the soil almost always provides direct economic benefits in terms of yield and production. Perhaps the major problem with such a conservation strategy is that it requires continuous, sensitive, and knowledgeable management of both the soil and the crop to be fully effective. But the rewards in terms of reduced soil loss are indisputable,[1] and it is a goal achievable by small or large, rich or poor farmers alike (39).

How vegetative cover and management work

Vegetative cover and rainfall. First and foremost, vegetation protects the soil from erosion by intercepting raindrops and absorbing

---

[1]It is not my purpose here to review the evidence for the efficacy of vegetative cover and management. Suffice to say, the differences in erosion between good cover/high management plots and clean-tilled, fallow plots can be in orders of magnitude of 100 or more. The reader is referred to Hudson (19), Meyer and Mannering (27), and the many other experiments worldwide.

their kinetic energies harmlessly. Some water may be evaporated from the leaves, but most reaches the ground surface either by stemflow or by reforming into droplets that, if the vegetative cover is close-growing, have little chance to pick up speed and gain further kinetic energy.

The German Ewold Wollny pioneered research on rainfall interception (47). He demonstrated how planting densities common for many crops allowed nearly 90 percent of the total annual rainfall to reach the ground unimpeded. Even under the highest planting densities, nearly half the rainfall in Wollny's experiments fell directly onto the soil (Table 1).

Although this has been a neglected area of research, it is clear from these early experiments and a few later ones, such as that by Sreenivas and associates (38), that raindrop interception is the main way that vegetation reduces erosion. Yet, if this were the whole story, a linear relationship would hold between percentage cover and erosion; it does not. Experimental evidence now indicates that the erosion-cover relationship is curvilinear and that erosion is little different whether cover is 100 percent or 60 percent (Figure 1). Researchers have found similar curvilinear relationships for runoff (16, 25). Such findings are vital to conservation research and planning in helping to design realistic management objectives. But they also raise the question as to why a mean seasonal cover of 60 percent is nearly as effective as full, continuous cover.

Vegetative cover and soil. There are many interactive processes between a plant and its soil that affect erosion. Some of these processes include the following:

▶ The physical binding of soil by plant stems and roots.

Table 1. Percentage of total rainfall penetrating a canopy of vegetation at different planting densities on 4-m² trial plots (30, 42, 47).

Planting Density (plants/m²)	Percent Total Annual Rainfall Penetrating Canopy With:			
	Corn	Soybeans	Oats	Peas
0	100	100	100	100
9	62.9	88.4	-	-
16	60.7	78.2	78.5	-
25	57.0	65.9	78.4	78.9
36	44.5	64.3	78.9	-

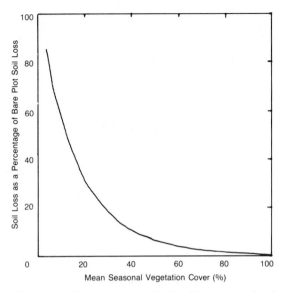

Figure 1. Erosion-cover relationship, generalized
after Elwell (13) and Elwell and Stocking (15).

▶ Electrochemical and nutrient bonding between roots and soil.
▶ Detention of runoff by stalks and organic litter.
▶ Improved infiltration along root channels.
▶ Greater incorporation of organic matter into the soil, resulting
in better structural and water-holding qualities.
▶ Increased faunal and biological activity, leading to better soil
structure.

Of course, many of these processes themselves contribute to better vegetative cover. This is particularly true in the seasonal tropics
where the increased infiltration promoted by vegetative cover helps
to avoid periodic drought stress and maintains a more vigorous cover.
Clearly, provided that vegetation is maintained above a certain level
of cover (50-60 percent, but varying according to type of cover and
soil), the interactive process between the soil and the plant are sufficient to cope with erosive forces.

Management. All of the above has direct implications for management. But in addition, management works to reduce erosion through
fertilization, timely planting, and a whole host of farm practices that

encourage vegetative growth. The only exception from normal routine farm practices is weeding, which inadvertently acts to reduce cover. Effects of management on soil characteristics are also important, especially in the maintenance of soil aggregation and the prevention of soil crusting. Aggregates are formed through the bridging effects of organic colloids. It is only through the use of mulches and other supplies of organic matter that aggregation can be kept stable.

Tillage practices are one of the main aspects of management that alter soil porosity and infiltration and incorporate organic residues into the topsoil. In particular, no-till practices and direct drilling techniques effectively reduce erosion, especially when combined with cover crops (20).

Ambivalent effects. It would be a mistake to call vegetative cover the panacea for soil erosion control. For example, a mature, natural forest cover is a perfect foil to erosion processes. Rates of erosion under undisturbed forest are usually always well below 500 kg/ha/yr. For example, in Hong Kong, on 29-degree slopes with pine regrowth, soil loss was 450 kg/ha (24). The temptation, therefore, is to promote afforestation as a universal conservation measure. However, I know of at least two cases where planted trees accelerated the erosion rate. In Brazil, on the slopes of the Val do Rio Doce, Minas Gerais, the planting of *Eucalyptus* spp. as an erosion control measure stifled ground cover and accelerated sheet erosion. In Mondoro, Zimbabwe, trees planted for gully control appear to have increased the rate of headward recession of gullies so that many of the trees have been undermined by the process they were meant to stop.

What are the causes of the ambivalent effects of vegetation? Three can be identified. First, the height of the vegetative cover above the ground surface is important (38). Droplets, often larger in mass than in the original rainfall, may reform on leaves. In falling to the ground, they may accelerate sufficiently to have a sizable kinetic energy. For example, a 2-mm diameter raindrop with a terminal velocity of 6 m/sec has about the same kinetic energy as a 3-mm drop falling at 3.25 m/sec, a speed that is reached well within the first meter of fall. Naturally, it requires more than three 2-mm drops to make a 3-mm one. Nevertheless, the energy impact of droplets falling from vegetation should not be underestimated.

Second, tall-growing vegetation may reduce ground cover completely, either directly by shading or by little-understood chemical effects through the roots of certain woody perennials. Unimpeded surface runoff, coupled with the bombshell effect of large droplets falling from leaves, can cause significant erosion.

Third, ambivalent effects of vegetation also have been demonstrated in laboratory rainfall simulator tests. De Ploey and associates (9) found a cover of grass definitely reduced erosion on slopes under 5 degrees. But above 8 degrees, the rate of erosion exceeded the rate on bare soil. The researchers concluded that higher slopes generate turbulent eddies downstream of grass blades that erode more soil. Obviously, there exists a complex interaction among vegetation, slope, soil type, and erosion.

Key areas in crop cover and management research

Before considering actual research methodologies, it is necessary to pinpoint the nature and type of research that may be needed in crop cover and management over the next few decades. For better or for worse, the greatest effort on a sustained, worldwide basis has gone toward establishing the C and P factor values in the universal soil loss equation. Many now feel that further refinement of such empirical methods through expensive research provides diminishing returns to knowledge and that scientists should pursue new areas of research more vigorously. Following are some areas that need attention.

Soil fertility, productivity, and management. One of the major lessons learned from recent erosion research is the complex nature of the interdependencies among erosion factors. In part, this is perhaps a reaction to the more simplistic erosion models, in which all factors are independent (indeed, this is a requirement of empirical-statistical models). Therefore, scientists increasingly have appreciated that the growth of good cover crops not only has an immediate benefit in reducing erosion but a carryover effect into succeeding years as well. For example, the antecedent effect of rotation is such that a maize-following-maize plot would always have higher soil losses than a maize-following-grass/legume plot. The classic analysis of Klemme and Coleman (21) on the differential effects of cropping systems on erosion is worth reading in this context. Similarly, it is a sad but true fact that areas which have suffered high erosion tend to continue to have high erosion rates. Some of the blame for this can be ascribed to erosion affecting fertility and fertility affecting crop growth. However, there is an enormous field of research open to investigation of such interactions. The Food and Agriculture Organization recently commissioned a study of erosion and productivity, and specific areas of research relevant to vegetation and management are identified in the report (41).

Quality of crop canopy. Vegetation varies significantly in its characteristics and structure, and this has a bearing on its efficacy for soil protection. Aspects might include the value of the vegetation as a mulch, the layout of leaves at different heights so that raindrop interception is maximized, good cover crops for low fertility/high erosion conditions (it is lamentable that nearly all experimental plot research is done under good, high-management conditions on soils that are uneroded), and density of planting as well as experiments with broadcast versus row versus grid planting patterns.

One line of research that needs investigation has to do with the role of stemflow. The structure of certain plants encourages water to quickly channel down the leaves and flow down the stem. Is this dangerous? De Ploey (8) demonstrated just how important stemflow is to runoff generation. Studies of the quality of the crop canopy could extend into crop breeding programs where the ability of a plant to produce a quick, close-growing canopy is as much a criterion for varietal selection as is yield and resistance to disease. In a similar vein, the development of cover crops that will yield a crop nearly every year under marginal, droughty, and low fertility conditions is important.

Economic aspects of speciality cover crops. One of the major factors responsible for the present high rates of erosion is continuous monocropping with no fallow or grass ley in the rotation. Farmers or landowners argue that it is not economical to have an unproductive year in a rotation and that inorganic fertilizers now make a recuperative year unnecessary. Sufficient evidence exists, however, that shows the deleterious effect of continuous cropping on soil structure and how more and more chemicals have to be applied simply to retain yields at present levels on many soils. More economic analyses are needed that take into account the dynamics of erosion/productivity relationships and incorporate the medium-term beneficial aspects of planted cover crops.

Integrated land use systems. Multiple cropping, agroforestry, alley cropping, interplanting, and stripcropping all have a certain appeal, not the least of which is the improved cover and soil conservation they afford. But the fact remains that, with notable exceptions, there is a dearth of experimental work on these systems, despite sometimes extravagant claims. In particular, there is little on-farm research into integrated land use. It is one matter to demonstrate that a cropping or management system is technically viable on a research station. It is quite another matter to do this under the real management constraints of a small farm. With modern farming techniques, the management of integrated forms of land use is difficult and requires specific skills on the part of the farmer. Yet, the scope is enormous for innovation and development of farming systems that are acceptable socioeconomically to the farmer while conserving the environment.

Agricultural operations on the farm. A particular case of farming systems research that has many implications for soil erosion is the timing of planting and harvesting. Planting date is critical to yield throughout the tropics. But it is also critical to the degree of protection afforded to the ground over a season by the growing crop. The later planting is delayed, the more opportunity rainfall has to hit bare soil. In the seasonal tropics especially, early season storms can be the most intense and erosive. Similarly, harvesting and disturbance of the soil can cause additional erosion. There are many and varied reasons for late planting. In developing countries, the most widely cited reason is the weakness of draught oxen at the end of the dry season to pull a plow through an often hard and stony soil. Whatever the precise cause, there is reason to examine farming systems with the view to proposing adaptations that would be beneficial to conservation objectives, for example, perhaps a breeding program for improved draught animals or fodder trees.

Tillage and soil structure research. Management also relates to appropriate methods of tillage and maintenance of soil structure. Type, direction, and degree of tillage all have important effects on erosion. There are indications that the same method of tillage year after year is bad for soil structure and that a rotation of implements is desirable, perhaps, in successive years, conventional moldboard plowing followed by zero tillage and direct drilling, light disking, and, finally, chisel plowing. Also, changes in soil properties following agricultural operations have a considerable effect on erodibility.

Methodologies

The previous discussion highlighted the fact that research on vegetation and management can be extremely diverse in scope and nature, varying from the micromorphological structure of plants, through plot and field experiments, to socioeconomic influences in farming systems. Clearly, it is impossible to detail all methodologies here. What follows is a description of only those approaches most directly relevant to vegetation, management, and erosion.

Vegetative cover. Botanists and ecologists tend to use basal cover—the area of the actual ground surface taken up by stalks of vegetation—as the principal measure of cover. Such measurement methods include line-intercept, wheel-point, and variable-plot techniques (44). For purposes of erosion research, however, it is more important to know the proportion of the ground that is covered, even if that cover is some distance off the ground. This gives a measure of the efficiency of the vegetation to intercept raindrops or, alternatively, the proportion of bare ground open to direct raindrop splash.

Techniques. The most simple, least costly, and most practical techniques for direct measurement of vegetative cover are ground-based, vertical photography and use of a quadrat sighting frame. Elwell and Gardner (18) evaluated the two techniques, concluding that the sighting frame is the better method for routine field measurements. Aerial photography is useful for resource surveys, but cannot provide the detail necessary for cover determinations; it is also a costly technique for repeated observations throughout a growing season.

Several authors have proposed stereophotography for monitoring vegetative changes (45). But this form of ground-based, vertical photography from a camera perched on a long pole suffers from excessive radial displacement. Even when used stereoscopically, the percentage of cover is always underestimated because it appears to be spread over a larger area than it actually is. Although Elwell and Gardner (18) worked out percentage errors for leaf canopies at different heights, they concluded that even with a camera located more than 5 m above the ground little more than immediate ground cover can be measured.

In its most simple form, a quadrat sighting frame consists of two horizontal bars set directly above each other. Ten small holes are drilled at regular intervals along each bar so that an observer may peer through a hole in the top bar and see a small area of the ground through the corresponding hole in the lower bar. The observer simply records the presence or absence of a leaf or other item of intercepting vegetation. After a predetermined number of observations, perhaps 100 sights or 10 positions of the frame for a fairly regular cover, or more if cover is discontinuous, the results are expressed as a percentage of hits in the total number of observations. An early user of this method estimated that accuracy with 10 random positions of the frame was ± 2 percent of vegetative cover (4). However, subsequent experience with the frame indicates that errors are underestimated seriously. Some sources of error include bias in the case of row crops, difficulties in vertical sighting with crops more than 1 m high, problems with partially covered sights, and difficulty of interpretation with heavy shadow. Unpublished tests using three observers on the same field showed that observer error and/or misinterpretation can produce unacceptable errors of ± 10 percent. Using similar procedures and statistical analysis, Elwell and Gardner (18) found that a total of 1,000 sights would be needed to achieve a 2 percent accuracy for uniform cover conditions and 300 sights for a 5 percent accuracy where cover is variable.

An improved instrument has been designed for vegetative cover measurements (Figures 2 and 3). It retains the ease of operation of the simple sighting frame but removes some of the biases and potential errors and allows the measurement of tall-growing vegetation, such as maize (14). Instead of looking vertically down, the new instrument uses adapted gunsights on sliding cursors (to allow adjustment for row crops at different spacings) to look obliquely downward onto a strip of mirror. The observer sees the reflected image of the crop leaves outlined against the sky.

Other techniques for assessing vegetative cover include the use of light meters, point quadrats, shadow measures (2), and electronic monitoring devices. One interesting cover study compared botanical methods, including all-contacts-point, first-contact-point, and step-point methods, evaluated against both runoff and sediment production in Colorado; none of the methods gave a good correlation against sediment yield (3). Indirect measures include assessing the effectiveness of cover in preventing soil detachment through the use of splash

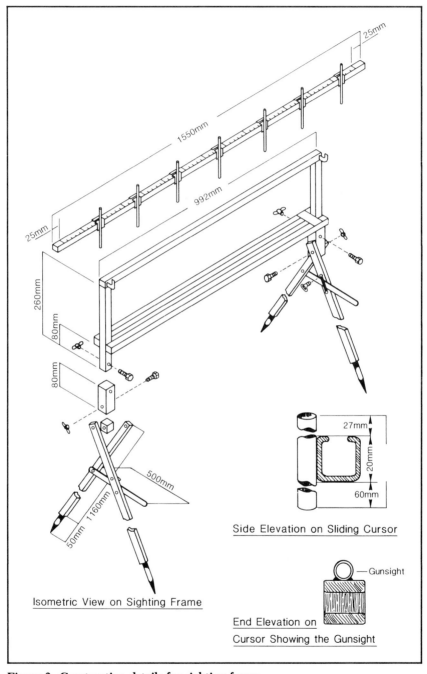

Side Elevation on Sliding Cursor

Gunsight

Isometric View on Sighting Frame

End Elevation on
Cursor Showing the Gunsight

Figure 2. Construction details for sighting frame.

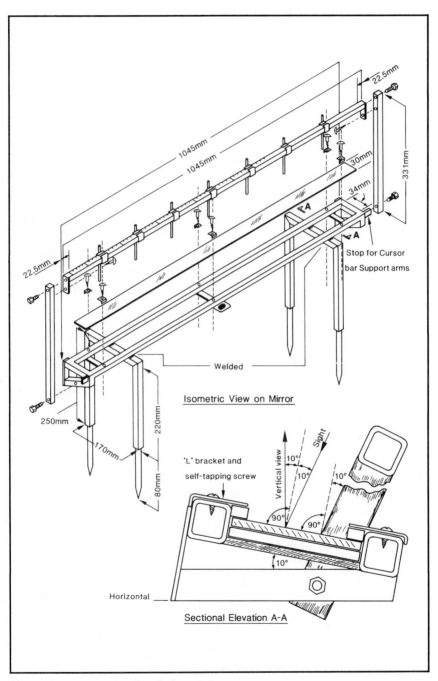

Isometric View on Mirror

Sectional Elevation A-A

Figure 3. Construction details for mirror.

cups. Sreenivas and associates (38) proposed the use of a soil cover rating as follows:

$$\text{SCR} = \frac{\text{weight of soil splashed from bare land} - \text{weight of soil splashed under cover}}{\text{weight of soil splashed from bare land}}$$

The technique has not received wide use, however.

Use of cover measurements. A measurement of cover percentage provides a single assessment at one point in time in the growth period of a crop or vegetation. In other words, if a rainfall event were to occur at that exact time, the cover value is equal to the percentage of rainfall likely to be intercepted before reaching the ground. By itself, such a measurement is of little intrinsic value. Knowledge is required of how the vegetation progresses through its growing season and the proportion of the seasonal rainfall that is intercepted. This seasonal interceptive efficiency of vegetation is one of the key parameters used in the soil loss estimation developed in and employed by the Conservation Service in Zimbabwe (17). It is similar to the crop protection factor of Sharma and associates (35).

To calculate seasonal interception, cover measurements are taken at regular intervals (usually every 10 days) through the growing season. Then a cover curve is constructed (Figure 4). Superimposed upon this curve is the mean seasonal distribution of rainfall kinetic energy as measured from a recording raingage. For example, if in one 10-day interval the cover averaged 25 percent and the kinetic energy of rainfall was 800 J/m², then 200 J/m² would be intercepted. Summing these calculations over the whole growth period, it is possible to calculate the percentage of mean seasonal kinetic energy that is intercepted by the vegetation. Typical values of interception range from about 20 percent for a poor crop of maize or tobacco to nearly 100 percent for a dense weed fallow.

Zimbabwe scientists have established a vegetative cover data bank of cover measurements and crop-hazard ratings. For instance, a cotton crop grown under moderate fertility and reasonable management will achieve 38 percent interception of rainfall, whereas a sorghum crop will achieve 52 percent interception. Researchers already know that the erosion-cover relationship is curvilinear (Figure 1) and that 38 percent cover is potentially dangerous. If erosion is likely to be severe because of slope or other conditions, then there would be compelling arguments to grow sorghum rather than cotton. The major variables affecting interceptive efficiency are planting date

and fertility status of the soil. In Zimbabwe, H. A. Elwell has constructed tables of interception for crops at various expected yield levels
and dates of crop emergence (Table 2 is an example for one crop).

Other vegetative characteristics and methods. Canopy height, the
structure of the plant, and rooting characteristics are measures of
vegetation that could influence erosion to varying degrees. Standard

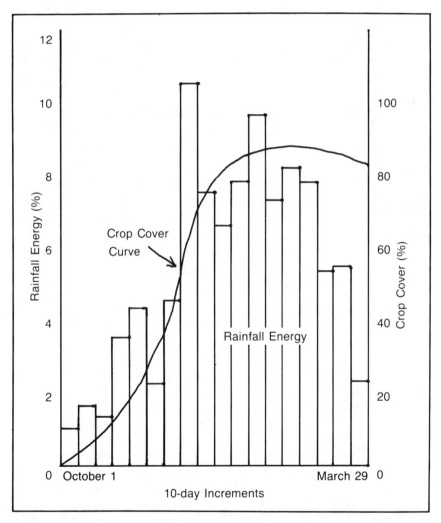

Figure 4. Crop cover and rainfall energy distribution (hypothesized) through a growing season, from which seasonal interception of rainfall is calculated.

length measures and a leaf area index can be used for the former, but root measurements are difficult. Because no single measure can adequately describe all vegetative characteristics that are important in erosion, one method is to classify vegetation into types based upon similarities in soil protective characteristics (Table 3).

The problem remains, however, to relate quantitatively types of vegetation to splash detachment and overall erosion. Laboratory and rainfall simulator studies have the most potential. Some possibilities that have been discussed at recent meetings, but which remain generally unpublished, include:

▶ Splash detachment in a rainfall simulator using plants grown in trays at various spacings and cover percentages: brussels sprouts, potatoes, and sugar beets have been used at Silsoe, England, for example.

▶ As above, but with plastic or artificial plants.

▶ As above, but using an open grid above the soil surface upon which discs or rectangles can be placed to simulate accurately various cover percentages and spacing of the cover. Experiments at the Institute of Agricultural Engineering, Harare, Zimbabwe, used this approach, along with clumps of *Hyparrhenia* grass to represent the stems of row crops (this addition of a basal cover reduces soil loss slightly but consistently).

The classic early experiments of Hudson (*19*) using mosquito gauze to simulate a full cover is another example.

Experiments on the effects of vegetative mulch on erosion are more common, principally because this is an aspect covered specifically by the USLE. Standard soil loss/runoff plot experiments (with or

Table 2. Interception values for cotton based upon yield and emergence data; data from H.A. Elwell, CONEX, Zimbabwe.

Expected Yield (kg/ha)	Interception Value (%) for Emergence Dates of:								
	1 Sept	15 Sept	1 Oct	15 Oct	1 Nov	15 Nov	1 Dec	1 Jan	1 Feb
500	43	41	39	34	29	23	16	7	2
1,000	62	59	55	49	41	32	24	9	2
1,500	72	69	65	57	48	38	28	11	2
2,500	84	79	75	66	56	44	32	13	3
3,500	92	87	82	72	61	48	35	14	3
4,500	95	89	84	74	63	49	36	14	3

without rainfall simulation) are the usual way of expressing the beneficial aspects of various rates of application and incorporation of organic residues and mulches (22, 36).

Plot and field experiments. Apart from experiments aimed at fixing factor values for the USLE (see later section), researchers widely use plot and field experiments to investigate specific crop and management circumstances. The major weakness of such experiments is this

Table 3. Crop classification based on similarities in soil-protective characteristics (16).

Description	Examples
A. Row crops	
1. Tall, upright crops generally grown on unridged lands	Annuals: maize, sorghum, sunflowers; perennials: napier fodder, sugarcane
2. Legumes, annuals; short, bunch and procumbent varieties	Beans: soya, velvet, etc.; groundnuts; cowpeas
3. Tall, upright crops on ridged lands	Tobacco; group 1 crops on ridges
4. Woody, bushy row crops	Cotton
B. Broadcast crops	
1. Tall, upright, for fodder	See A1
2. Short legumes broadcast for fodder and green manure	See A2
3. Medium height plants for fodder, green manure, etc.	Sunhemp, weed fallow
C. Orchards/plantations	
1. Individual trees and bushes planted on regular pattern	Coffee, citrus, deciduous fruit
2. Hedged crops	Tea
3. Thick stands of natural and exotic trees with little to no grass cover	Forestry
D. Grasslands	
1. Stoloniferous grasses planted in in rows from runners; permanent pastures	Star, Kikuyu, torpedo
2. Seed established grass; bunch varieties	Love grass, *Sabi panicum*, Rhodes, etc.
3. Species composition related to natural soil and environmental conditions	Natural veld grasses, mixed species, mainly bunch, with some annuals and perennials

very specificity; they are only relevant to the precise conditions of the experiment. Extrapolation must proceed with extreme caution. A few of the many examples that might be quoted include the following:

► Water yield and erosion response to land management on watersheds up to 389 acres (37).

► Comparison of the effects of no-till practices and conventional plowing on soil properties and yields of different crops. No-till had negligible soil loss (23).

► Effect of narrower row spacing for a sorghum crop. Soil loss was 39 percent less than with conventional spacing (1).

All experiments that monitor the relationships among erosion and measures of vegetation and management show that other factors influence the exact nature of the relationship. Some of these factors are the nature of the rainfall, evapotranspiration, and infiltration rates. In turn, these factors may depend at least partially on the vegetation and management. For the researcher there is often the considerable danger of circularity of argument, as well as lack of adequate control over plot and field experiments.

Vegetation as an indicator of erosion. Within the complexity of natural systems, one variable effectively can integrate several other variables and present an observer with useful information in a simplified manner (43). Plant indicators are one of the most powerful integrators, and they have two important areas of application in erosion research: in general survey work for use as dateable markers of old soil levels and through species composition as indicators of the state of the environment or of soil types or conditions that are particularly susceptible to erosion.

Field measurements using vegetation. Dunne and associates (10) described several ways of using vegetation as a rapid means of field measurement of erosion. The exposure of tree roots is one common method. But a more accurate assessment used in Tanzania involves use of certain woody species, for example, ant-gall acacia (*Acacia drepanalobium*), that have a physiological mark on their stem (a bulge or branching or color of bark) at the level of the original soil surface (11). With suitable species, it is possible to develop a calibration curve between age of tree (by ring count) and circumference at ground level. Therefore, nondestructive sampling can be carried out linking age of tree and depth of erosion since germination of the tree. There are several practical difficulties leading to sources of error with this methodology, such as ambiguity of original surface level or scarcity of suitable shrubs in some places. But it is an unrivalled method for gaining recent historical data on rates of sheet erosion without the need for expensive instrumentation.

Indicator species. As indicators, plants are extremely useful because they are obvious features of the landscape and can react dramatically to small changes in environmental conditions. The reaction of grasslands, in particular, to overgrazing and intensive use is usually by a sequential change in species composition from vigorous, palatable perennial grasses to a reduced density of tufts of less palatable annual species (*32*). In central Zimbabwe, I observed *Panicum coloratum* with some *Digitaria* spp. on a lightly grazed site and *Brachiaria humidicola* on an adjacent but intensively used site— an overall serious deterioration of the grassland that can be used as an indirect measure of erosion.

Certain species also tend to be specific to particular soils. Apart from the obvious role of plant indicators for soil surveys, plants can be used to identify certain problem soils from the standpoint of erosion. Throughout central Africa, the mopane tree (*Colophospermum mopane*) is an immediate indicator of sodic soils that are extremely erodible. A less well-known example is the occurrence of mupundu (*Parinari curatellaefolia*), which flourishes on ridge tops in zones with a fluctuating seasonal water table in the root zone. Such conditions form noncalcic hydromorphic soils that have an open-bridge structure, strong when dry but susceptible to sudden collapse when wet. With the use of agricultural machinery, the management of such a soil is especially critical.

Cropping factors in predictive equations. Two USLE factors relate directly to vegetation and management. The cover and management factor, C, is the ratio of soil loss from an area with specified cover and management to that from an identical area in tilled, continuous fallow. The support practice factor, P, is the ratio of soil loss with a support practice, such as contouring, stripcropping, or terracing, to that with straight-row farming up-and-down slope. Field plot measurements form the basis for determining these factor values. Specific procedures for C and P values are laid down in the Agriculture Handbook 537 (*46*).

Briefly, C values for a particular site are calculated from soil loss ratios representing six cropstage periods (rough fallow, seedbed, establishment, development, maturing crop, and residue/stubble) and three levels of canopy cover at the mature stage. Other variables include crop type; the specific rotation; stage in the rotation; method of plowing—moldboard, conventional, chisel; direction of plowing; use of winter cover crops; dry weight of residues in the spring; effect of chopping stalks for residue; and percentage of soil surface covered by mulch after crop seeding.

Some crops, for example, require further refinements and details to calculate accurate C values. A large number of interrelated variables are covered by this one factor. Consequently, there has to be considerable complexity both in method of determination and in final C values. For example, maize has some 60 C values in the United States to account for the variety of ways it is grown.

The P factor basically describes the effect of those management practices that are conservation-oriented: contouring, ridge planting, stripcropping, and terracing. It is more simple in its application than C, but there is considerable overlap in concept between C and P, with improved tillage, sod-based rotations, fertility treatments, and crop residues relegated to C rather than P.

Considerable research has gone into determining USLE factor values. Noteworthy in this respect are the research programs funded by the French Office for Overseas Scientific Research (ORSTOM) in West Africa and Brazil (26, 33). It remains an unfortunate fact that the complexity of tropical farming systems necessitates a vastly greater research effort to establish C and P values than has already been conducted in the United States, an impossible task with present resource constraints. Perhaps for individual high-value crops, the research effort may be worthwhile.

Other models tend not to address the complexities of vegetation and management. One exception is SLEMSA (Soil Loss Estimator for Southern Africa), which uses the concept of rainfall interception described earlier. It also incorporates management effects into soil erodibility on the basis that plowing, residues, and other standard management practices basically affect the susceptibility of the soil to erosion. This is one of the integrative components of SLEMSA that makes it somewhat different than the USLE.

Farming systems approaches. Farming systems research is now seen as a research methodology in its own right, although it has only rarely been applied specifically to soil conservation (40). If the objectives of soil conservation are to save soil and maintain productivity, then there are many means to that end. The most obvious and widely used means is to implement physical conservation works and land use planning. But time and again these have failed. What are the alternatives? At least five separate strategies have been identified (39), but all of them demand an intimate knowledge of farm practices, social and economic circumstances, labor peaks, and the like before they can be implemented with any confidence that they will be accepted. Farming systems research aims to analyze the farming system background to pinpoint any positive opportunities within the day-to-day life of the farm household in which practices that also meet soil conservation objectives can be incorporated—it is "conservation by stealth"! Notwithstanding the emphasis at some international research centers, farming systems research is not simply a technical analysis of cropping systems.

A key manual on farming systems research methodology is *Planning Technologies Appropriate to Farmers: Concepts and Procedures* (6), published by CIMMYT (Centro International de Mejoramiento de Maiz y Trigo). It emphasizes a point that is particularly relevant for soil conservation research and planning:

"Few farmers are following in their entirety the recommendations made by researchers and extension workers. Some argue that farmers are at fault, some that extension is ineffective, others that credit is unsuitable, and some that inputs are not available in a timely way. A less frequently heard explanation is that the recommended technologies themselves are simply not appropriate to farmers."

Adoption of a technically proven method of soil conservation by farmers is, in effect, the same procedure as the adoption of any new technology. That adoption hinges upon a number of interrelated factors. In the main, farmers seek technologies that increase their incomes while keeping risks within reasonable bounds. Whether one talks to a rich commercial farmer in a developed country or a shifting cultivator in the tropics, it is of little practical use to appeal to his or her conservation ethics. Farming systems research, in analyzing the interrelationships in decision-making on the farm, hopefully may identify where new technology and what type of technology can be introduced without upsetting the delicate set of conditions that the farmer perceives to be important.

CIMMYT (6, 29, 34) has identified a systematic set of procedures that are useful in obtaining information on a farmer's circumstances. In brief, the procedure involves:

▶ Assembling background information, such as agro-climatic, erosion, population data, etc., from published and unpublished secondary sources.

▶ Making exploratory surveys—informal interviews with farmers, extension workers, merchants, and others with strong local knowledge as a means of narrowing down the circumstances of farmers. A guiding principle adopted by most farming systems research workers is that if a significant number of farmers in a region are using (or not using) a particular practice there is a good reason.

▶ Making a formal survey—formal interviews by trained interviewers using specific and focussed questions identified in the exploratory survey but needing qualification and verification.

A closely related research approach emphasizing vegetation and management is that adopted by agroforestry research. One of the principal objectives of agroforestry is soil conservation and the sustainability of agricultural/forestry/livestock production in difficult, erosion-vulnerable environments. The International Council for Research in Agroforestry, based in Nairobi, Kenya, has developed "A Diagnostic and Design Methodology for Agroforestry" (31) that incorporates elements of farming systems research in the context of rapid rural appraisal to identify the possibilities for the growth of woody plants in conjunction with other land use activities. Indeed, "rapid rural appraisal" (5), which has entered into the jargon of tropical development work, is a set of methodologies and techniques designed to maximize the social relevance of project planning activities without resorting to long, time-consuming, expensive survey work—an application especially relevant to soil conservation planning (7). Useful listings of plants suitable for soil erosion control and agroforestry purposes have been developed (12, 28).

Farming systems research involves a diversity of individual approaches. Of late, there has been a tremendous increase of interest in this line of investigation. If there is one lesson to be drawn from farming systems research that is relevant to all methodologies for the assessment of vegetative cover and management, it is that techniques for research and for planning in soil conservation need to be adapted far more closely to the farmer and his or her local environment. Universal solutions are impossible. The challenge for research is how best to identify human and physical environmental conditions at the local level rapidly and accurately enough to plan conservation effectively.

REFERENCES

1. Adams, J. E., C. W. Richardson, and E. Burnett. 1978. *Influence of row spacing of grain sorghum on ground cover, runoff and erosion.* Soil Science Society of America Journal 42: 959-962.
2. Adams, J. E., and G. R. Arkin. 1977. *A light interception method for measuring row crop ground cover.* Soil Science Society of America Journal 41: 789-792.
3. Branson, F. A., and J. E. Owen. 1976. *Plant cover, runoff and sediment yield relationships on Mancos shale in western Colorado.* Water Resources Research 6: 783-790.
4. Cackett, K. E. 1964. *A simple device for measuring canopy cover.* Rhodesian Journal of Agricultural Research 2(1): 56-57.
5. Chambers, R. 1981. *Rapid rural appraisal: rationale and repetoire.* Public Administration and Development 1: 95-106.
6. Centro Internacional de Mejoramiento de Maiz y Trigo. 1980. *Planning technologies appropriate to farmers: concepts and procedures.* El Batan, Mexico.
7. Collinson, M. P. 1981. *A low cost approach to understanding small farmers.* Agricultural Administration 8(6): 433-456.
8. De Ploey, J. 1982. *A stemflow equation for grasses and similar vegetation.* Catena 9: 139-152.
9. De Ploey, J., J. Savat, and J. Moeyersons. 1976. *The differential impact of some soil loss factors on flow, runoff creep and rainwash.* Earth Surface Processes 1: 151-161.
10. Dunne, T., W. E. Dietrich, and M. J. Brunengo. 1978. *Recent and past erosion rates in semi-arid Kenya.* Zeitschrift for Geomorphologie Supplementband 29: 130-140.
11. Ecosystems, Ltd. 1982. *Southeast Shinyanga Land Use Study. Report to World Bank/Shinyanga RIDEP/Government of Tanzania.* EcoSystems Ltd., Nairobi, Kenya.
12. El-Swaify, S. A., E. W. Dangler, and C. L. Armstrong. 1982. *Soil erosion by water in the tropics.* Research Extension Series 024. College of Tropical Agriculture and Human Resources, University of Hawaii, Manoa.
13. Elwell, H. A. 1980. *Design of safe rotational systems.* Department of Conservation and Extension, Harare, Zimbabwe. 50 pp.
14. Elwell, H. A., and F. Wendelaar. 1977. *To initiate a vegetal cover data bank for soil loss estimation.* Research Bulletin 23. Department of Conservation and Extension, Harare, Zimbabwe.
15. Elwell, H. A., and M. A. Stocking. 1974. *Rainfall parameters and a cover model to predict runoff and soil loss from grazing trials in the Rhodesian sandveld.* Proceedings of the Grassland Society of South Africa 9: 157-164.
16. Elwell, H. A., and M. A. Stocking. 1976. *Vegetal cover to estimate soil erosion hazard in Rhodesia.* Geoderma 15: 61-70.
17. Elwell, H. A., and M. A. Stocking. 1982. *Developing a simple yet practical method of soil loss estimation.* Tropical Agriculture (Trinidad) 59(1): 43-48.
18. Elwell, H. A., and S. Gardner. 1975. *Comparison of two techniques for measuring percent canopy cover of row crops in erosion research programmes.* Research Bulletin 19. Department of Conservation and Extension, Harare, Zimbabwe.

19. Hudson, N. W. 1957. *Erosion control research. Progress report on experiments at Henderson Research Station, 1953-56.* Rhodesian Agricultural Journal 54(4): 297-323.
20. Kemper, B., and R. Derpsch. 1981. *Results of studies made in 1978 and 1979 to control erosion by cover crops and no-tillage techniques in Parana, Brazil.* Soil and Tillage Research 1: 253-267.
21. Klemme, A. W., and O. T. Coleman. 1949. *Evaluating annual changes in soil productivity.* Bulletin 522. Missouri Agricultural Experiment Station, Columbia.
22. Kramer, L. A., and L. D. Meyer. 1969. *Small amounts of surface mulch reduce soil erosion and runoff velocity.* Transactions, American Society of Agricultural Engineers 12(5): 638-641, 645.
23. Lal, R. 1976. *No tillage effects on soil properties under different crops in Western Nigeria.* Soil Science Society of America Journal 48: 762-768.
24. Lam, K. C. 1978. *Soil erosion, suspended sediment and solute production in three Hong Kong catchments.* Journal of Tropical Geography 47: 51-62.
25. Lang, R. D. 1979. *The effect of ground cover on surface runoff from experimental plots.* Journal of the Soil Conservation Service of New South Wales 35(2): 100-114.
26. Leprun, J.-C. 1981. *A erosao, a conservacao e o manejo de solo no nordeste Brasileiro: balanco, diagnostico e novas linhas de pesquisas. Recursos de Solos, SUDENE-DRN, 15.* Superintendencia do Desenvolvimento do Nordeste. Recife, Brazil.
27. Meyer, L. D., and J. V. Mannering. 1971. *The influence of vegetation and vegetative mulches on soil erosion.* In *Biological Effects in the Hydrological Cycle.* Proceedings, Third International Seminar for Hydrology Professors. Purdue University, West Lafayette, Indiana. pp. 355-366.
28. Nair, P.K.R. 1980. *Agroforestry species. A crop sheets manual.* International Council for Research in Agroforestry, Nairobi, Kenya.
29. Norman, L. W. 1980. *The farming systems approach: Relevancy for the small farmer.* Rural Development Paper No. 5. Department of Agricultural Economics, Michigan State University, East Lansing.
30. Peterson, J. B. 1964. *The relation of soil fertility to soil erosion.* Journal of Soil and Water Conservation 19: 15-19.
31. Raintree, J. B. 1982. *A methodology for diagnosis and design of agroforestry land management systems.* International Council for Research in Agroforestry, Nairobi, Kenya.
32. Roberts, B. R., E. R. Anderson, and J. M. Fourie. 1975. *Evaluation of natural pastures; quantitative criteria for assessing condition in the* Themeda *veld of the Orange Free State.* Proceedings, Grassland Society of South Africa 10: 133-140.
33. Roose, E. J. 1977. *Erosion et ruisellement en Afrique de l'ouest. Vingt annees de mesures en petites parcelles experimentales.* ORSTOM, Paris, France.
34. Shaner, W. W., P. P. Philipp, and W. R. Schmel. 1982. *Farming systems research and development.* Westview Press, Boulder, Colorado.
35. Sharma, S. S., R. N. Gupta, and K. S. Panwar. 1976. *Concept of crop protection factor evaluation in soil erosion.* Indian Journal of Agricultural Research 10(3): 145-152.
36. Soil Conservation Society of America. 1979. *Effects of tillage and crop residue removal on erosion, runoff and plant nutrients.* Special Publication No. 25, Ankeny, Iowa.
37. Spomer, R. G., K. E. Saxton, and H. G. Heinemann. 1973. *Water yield and*

erosion response to land management. Journal of Soil and Water Conservation 28(4): 168-171.
38. Sreenivas, L., J. R. Johnston, and H. W. Hill. 1947. *Some relationships of vegetation and soil detachment in the erosion process.* Proceedings, Soil Science Society of America 12: 471-474.
39. Stocking, M. A. 1983. *Development projects for the small farmer: Lessons from east and central Africa in adopting soil conservation.* In *Soil Erosion and Conservation.* Soil Conservation Society of America, Ankeny, Iowa, pp. 747-758.
40. Stocking, M. A. 1983. *Farming and environmental degradation in Zambia: The human dimension.* Applied Geography 3: 63-77.
41. Stocking, M. A. 1984. *Erosion and soil productivity: A review.* FAO Consultants' Working Paper No. 1. Soil Conservation Programme, Land and Water Development Division, Food and Agriculture Organization, Rome, Italy.
42. Stocking, M. A. and H. A. Elwell. 1976. *Vegetation and erosion: A review.* Scottish Geographical Magazine 92(1): 4-16.
43. Stocking, M. A., and N.O.J. Abel. 1981. *Ecological and environmental indicators for the rapid appraisal of natural resources.* Agricultural Administration 8: 473-484.
44. Walker, B. H. 1978. *An evaluation of eight methods of botanical analysis on grassland.* Journal of Applied Ecology 7: 403-416.
45. Wells, K. F. 1971. *Measuring vegetation changes on fixed quadrats by vertical ground stereophotography.* Journal of Range Management 24(3): 233-235.
46. Wischmeier, W. H., and D. D. Smith. 1978. *Predicting rainfall erosion losses— a guide to conservation planning.* Agriculture Handbook No. 537. U.S. Department of Agriculture, Washington, D.C..
47. Wollny, E. 1890. *Untersuchungen uber das Verhalten der atmospharischen Niederschlage zur Pflanze und zum Boden.* Forsch. Geb. Agriphys. 13: 316-365. (Reported by Baver, L. D. 1938. *Ewald Wollny—a pioneer in soil and water conservation research.* Proceedings Soil Science Society of America 3: 330-333).

9
R. Lal

Monitoring soil erosion's impact on crop productivity

QUANTIFYING the effects of soil erosion on crop yields is a complex task because it involves the assessment of a series of interactions among soil properties, crop characteristics, and the prevailing climate. The effects are also cumulative and often not observed until long after accelerated soil erosion begins. Furthermore, the magnitude of erosion's effects on crop yields depends upon soil profile characteristics and on management systems. Crop yield, an integrated response to many interacting parameters, is difficult to relate under field conditions to any individual factor. It is, therefore, difficult to establish a one-to-one, cause-and-effect relationship between rates of soil erosion and erosion-induced soil degradation on the one hand and crop yields on the other.

Nonetheless, it is imperative that the erosion-productivity relationship be known for major soils. Such information is essential for future planning and for developing an effective land use policy. The customary method of reporting soil erosion—the equivalent depth per unit time or equivalent mass of soil displaced per unit time—should be replaced by an expression of economic loss in monetary terms. The economic loss attributable to soil erosion includes loss of applied and inherent plant nutrients, including soil organic matter reserves, loss of plant-available water reserves and storage capacity, crop burial, and stand losses. In addition to reduced grain yield, erosion also increases crop production costs. Improved technology often masks the effects of lost fertility and water storage capacity, making these effects difficult to quantify.

Because of pollution of surface water and groundwater and the

187

rising costs of agricultural inputs, there has been an increased interest in soil erosion and erosion-caused productivity decline (2, 7, 13).

Evidence and causes of yield decline due to soil erosion

Stocking (21) attributed the lack of a direct relationship between soil erosion and productivity to four causes:

► Soil erosion and productivity, while interdependent, are not independent variables.

► Under field conditions, reduced crop yield often is attributed to factors other than those related to soil erosion, such as pests, climatic change, soil salinity, water-logging, compaction, or others.

► Differences in soil profile characteristics cause differential effects on yield in soils with similar levels of soil erosion.

► Most parameters that affect crop yields and are affected by soil erosion are interrelated. A change in one variable induces changes in others.

Soil erosion influences crop yield by changing factors that limit production. In other words, progressive soil erosion increases the magnitude of soil-related constraints to production. These constraints can be physical, chemical, or biological. Among the important soil physical constraints aggravated by erosion are reduced rooting depth, loss of soil water storage capacity, crusting and soil compaction, and hardening of plinthite. Erosion also changes soil color and albedo. Moreover, erosion results in loss of clay and colloids due to preferential removal of fine particles from the soil surface. The loss of clay influences soil tilth and consistency. Exposed subsoil is often of massive structure and harder consistency than the aggregated surface soil. Development of rills and gullies may change microrelief, create soil variability, and render mechanized farm operations difficult. Another physical effect of soil erosion concerns the management and timing of farm operations. Achieving a desired seedbed with friable tilth necessitates a delay in plowing until the soil is adequately wet. Soil chemical constraints and nutritional disorders related to erosion include low cation exchange capacity, deficiency of major plant nutrients (N,P,K) and trace elements (Zn, S), nutrient toxicity (Al, Mn), and high soil acidity. Soil biological properties, important factors related to productivity, include biomass carbon and activity of macrofauna, such as earthworms. Erosion-induced alterations affect soil biological properties.

The loss in productivity set in motion by accelerated soil erosion is a self-sustaining process. Loss of production on eroded soil further degrades its productivity, which, in turn, accelerates soil erosion. The cumulative effect observed over a long period of time may lead to irreversible loss of productivity in shallow soils with hardened plinthite or in soils that respond only to expensive management and to additional inputs.

Monitoring erosion's impacts on crop production

In 1984 a group of scientists recommended possible field, greenhouse, and laboratory methods to evaluate the effects of soil erosion or productivity (21). Numerous techniques were suggested to establish the cause-and-effect relationship between erosion and yield. These were grouped into four categories: agronomic or direct methods, assessment of soil properties, geological measurements of weathering rate, and modeling and productivity indexes.

Direct methods. Productivity losses can be measured under field conditions where crop performance over a period of time is related to recorded soil loss or to erosion-induced alterations in soil properties. Direct methods comprising agronomic experiments with natural or artificial soil loss are described below.

Yield records from long-term agronomic trials. Evaluation of long-term yield records of agronomic experiments conducted on different soils can provide some indirect measure of changes in productivity due to changes in soil properties. For example, response functions over time for soils A to D in figure 1 indicate differences in yield reductions due to alterations in soil properties. Crop yield on soil A apparently was affected more drastically than on soils B, C, or D. Although agronomic records of yield are available, it is often difficult to evaluate soil properties and attribute the magnitude of changes caused by erosion.

Yield trends over time also can be assessed for different soil surface management treatments over the same soil. The yield data in figure 1 could be obtained from a tillage experiment conducted on sloping land. The differential yield trends for treatments A and B may be related to effects of different levels of erosion caused by different tillage methods. Such records from well-planned tillage experiments conducted for a reasonably long period are rather rare.

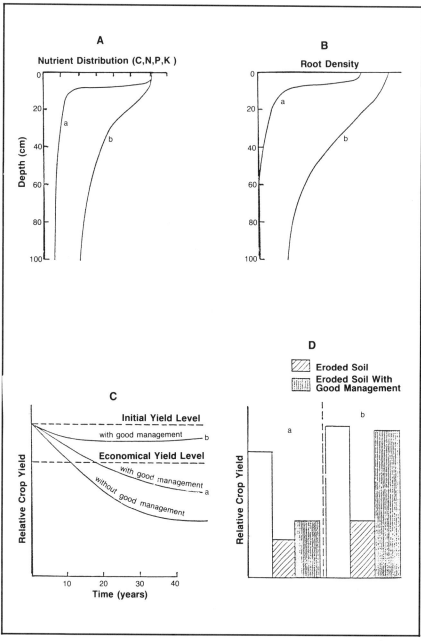

Figure 1. Crop response to erosion in relation to soil profile characteristics for nutrient distribution and rooting density. Yield trends with time in C and D may be attributed to different levels of erosion caused by tillage systems.

Table 1. Relation between soil erosion, as measured on field runoff plots from 1970 to 1975, and grain yields of maize and cowpea on an Alfisol in southwestern Nigeria (11).

Slope (%)	Regression Equation	Correlation Coefficient
Cowpea		
1	$Y = 0.43 \exp(-0.036X)^*$	-0.85^A†
5	$Y = 0.64 \exp(-0.06X)$	-0.97^B
10	$Y = 0.49 \exp(-0.004X)$	-0.91^A
15	$Y = 0.29 \exp(-0.002X)$	-0.66
Maize		
1	$Y = 6.41 \exp(-0.017X)$	-0.99^C
5	$Y = 6.70 \exp(-0.003X)$	-0.99^C
10	$Y = 6.70 \exp(-0.003X)$	-0.89^B
15	$Y = 8.36 \exp(-0.004X)$	-0.86^A

*Y is grain yield in ton/ha; X is cumulative soil erosion in ton/ha.
†A = significant at 5% level of probability; B = significant at 1% level of probability; C = significant at 0.1% level of probability.

Because experiments of this nature are not designed to evaluate the effects of erosion, such information is of an indirect nature. These experiments also lack a control over changes in soil properties. Careful evaluation of the analytical data on soil properties from different treatments may provide clues to the control variables that influence crop yield, such as available water storage capacity, nutrient status, or rooting depth.

Using long-term erosion plots for agronomic experiments with known soil loss. Another possibility is to conduct long-term crop yield assessment experiments with standard management where plot history is known and adequately recorded. These plots may be field runoff plots on which past erosion has been recorded precisely. Effects of past erosion on crop yields can be obtained by conducting agronomic experiments with uniform or variable management. I used such a procedure to monitor soil erosion on field runoff plots from 1971 to 1975 under different soil surface management systems on an Alfisol near Ibadan, Nigeria (11). Maize and cowpeas were grown in 1976 and 1977 as sequential crops under uniform soil management systems. The effects of previous soil erosion on different plots were related to maize and cowpea yield. Table 1 shows that maize and cowpea yields declined exponentially with cumulative increases in soil erosion. The high correlation coefficient indicates the strong influence of erosion-induced changes in soil properties. Erosion-caused productivity losses may be less for soils with deep, effective rooting depth and with edaphologically favorable subsoil characteristics.

New erosion plots under natural or simulated rainfall. This is the most direct method to assess the effects of erosion on yield because the experiment is designed for this specific objective. Accurate records of soil loss and changes in soil properties are maintained over a period of time. Different levels of natural soil erosion over time are achieved by varying surface soil conditions, for example, plowed bared, natural vegetative cover, plowed but with a protective inorganic screen mulch cover, etc. The erosion process can be accelerated by applying simulated rainstorms under field conditions. Large field-type simulators are expensive, however, and produce variable levels of soil erosion. Appropriate crops are then grown with uniform soil and crop management systems. Figure 2 is an example of field runoff plots with a runoff and soil loss collection system.

Figure 2. Field runoff plots showing details of a runoff collection system used to measure soil loss.

Table 2. Regression equations relating grain yields of maize and cowpea to simulated erosion by desurfacing for different soils and ecological regions in Nigeria (14).

Location	Soil	Annual Rainfall (mm)	Regression Equation	Correlation Coefficient R
Maize				
Onne	Oxic Paleudults	2,480	$Y = 1,590 - 233E + 7.68E^2$*	0.94[C]†
Ikenne	Oxic Paleustalfs	1,307	$Y = 11,869 - 725E + 12.1E^2$	0.82[B]
Ilora	Oxic Paleustalfs	1,230	$Y = 4,537 - 357E + 7.18E^2$	0.61[A]
S. Nigeria	Alfilsols & Ultisols	1,230-2,480	$Y = 5,999 - 435E + 9.00E^2$	0.43[A]
Cowpea				
Onne	Oxic Paleudults	2,480	$Y = 534 - 65.4E + 2.39E^2$	0.91[C]
Ikenne	Oxic Paleustalfs	1,307	$Y = 832 - 50.7E + 1.08E^2$	0.69[A]
Ilora	Oxic Paleustalfs	1,230	$Y = 1,522 - 65.3E + .144E^2$	0.77[B]
S. Nigeria	Alfilsols & Ultisols	1,230-2,480	$Y = 963 - 60.5E + 1.21E^2$	0.53[B]

*Y is grain yield in kg/ha; E is cm of soil removed.
†A = significant at 5% level of probability; B = significant at 1% level of probability; C = significant at 0.1% level of probability.

Desurfacing experiments. Artificial removal of surface soil to varying depths, followed by crop growth evaluation under uniform management, is a common agronomic technique (10, 11, 14, 15). The technique is rapid, simple, relatively inexpensive, but unnatural. Mbagwu and associates (14, 15) conducted desurfacing experiments in southern Nigeria and developed regression equations relating maize and cowpea yields to the depth of topsoil removed. Their results showed linear and quadratic effects of soil removal depth on maize and cowpea yields (Table 2). The effects of desurfacing were more severe on maize than on cowpea and, therefore, were soil- and crop-specific. I evaluated the effects of desurfacing of an Alfisol at the International Institute for Tropical Agriculture, near Ibadan, Nigeria, for varying levels of N and P (10). Application of major plant nutrients could not fully compensate for topsoil loss. It is possible that, in addition to the drastic reduction in available water, deficiency of some trace elements and toxicity of Al and/or Mn reduced grain yields.

Effects of desurfacing are different from natural soil erosion processes. Natural erosion is a sorting process involving preferential displacement of clay and organic colloids and removal of soil from some places and deposition at others. In contrast, desurfacing is a wholesale removal of a uniform soil depth. Consequently, the effects on crop growth and yields likely will differ. The effects of desurfacing likely are more drastic in soils where nutrient fertility is concentrated in the top few centimeters and subsoil horizons are edaphologically unfavorable. In comparison, the effects likely are less drastic on soils with deep surface soil and favorable subsoil properties.

I compared the effects of desurfacing and natural erosion on maize grain yield for an Alfisol at Ibadan, Nigeria (11). Maize grain yield fell 0.13 and 0.09 t ha^{-1} cm^{-1} of eroded soil for desurfacing to 10 and 20 cm depths, respectively. On the same soil about 10 m away, however, the decline in maize grain yield caused by natural erosion was 0.26 t ha^{-1} mm^{-1} of eroded soil. The yield reduction by natural erosion was about 16.25 times greater than that caused by artificial desurfacing. For soils where chemical fertility is confined to the top, thin soil layer, the effects on yield of removing 1 cm of surface soil are more drastic than the average effect spread over 10 cm of soil depth removed by desurfacing. It is important to note that the enrichment ratio of eroded sediments under natural erosion is 3:5 for organic matter, clay, and plant nutrients (10).

Laboratory and greenhouse studies. The relative crop response with varying inputs of nutrients and water and at varying levels of compaction can be obtained readily in greenhouse or growth chamber studies using soil samples from appropriate depths to simulate different levels of erosion. The effects of natural erosion can be simulated by sieving out the fine soil fraction. The crop growth information thus obtained, through of a relative nature, can provide basic information on control factors, such as drought, nutrient imbalance, etc. Aina and Egolum (1) used this method to evaluate maize response to surface and subsoil conditions. Their data indicated yield improvement in maize by adding N, P, and K and high doses (3 and 6 percent) of cattle manure. One limitation of this method is the artificial conditions of crop growth in a greenhouse environment on disturbed samples packed in a confined space, which, again, makes the data obtained of relative importance.

Assessment of soil properties. Two major effects of soil erosion are degradation of soil physical properties and loss of plant nutrients and soil organic matter. Rather than evaluating crop response agronomically on plots on which the surface soil loss has been measured by direct methods, researchers often estimate erosion indirectly from

Table 3. Influence of erosion on soil physical and chemical properties (8).

Soil Parameter	Regression Equation	Correlation Coefficient
Organic carbon (%)	Y = 1.79 – 0.002(E)*	– 0.71
Total nitrogen (%)	Y = 0.163 – 0.0002(E)	– 0.60
Bray-1 (ppm)	Y = 80.8 – 0.13(E)	– 0.77
Soil pH	Y = 5.6 – 0.02(E)	– 0.92
Total porosity	Y = 38.7 – 0.02(E)	– 0.92

*E is soil erosion (t ha^{-1})

changes in soil physical and chemical properties. Less accurate, indirect methods of estimating soil erosion are less capital-intensive, however.

Field surveys. Field surveys of soil profile characteristics are used to assess the approximate loss of surface soil. Field surveys can be supported by detailed laboratory analyses of soil chemical and physical properties to assess the depth of topsoil removed by past erosion. This paired-plot evaluation technique is useful provided that relatively uneroded and eroded plots exist side-by-side. While soil erosion is estimated indirectly from changes in soil properties over time, crop response is measured under similar agronomic practices of tillage, fertilizer application, and soil surface management.

One drawback of this indirect technique of estimating soil erosion is that changes in soil properties can occur with cultivation even without soil erosion. Disturbance of the soil surface by plowing and exposure to climatic elements causes clay migration to the subsoil horizon, reduced soil organic matter content, and other changes in soil properties similar to those caused by sheet erosion. Knowledge of plot history and previous land use is very important. I observed significant relationships between soil properties and the accumulative soil erosion (Table 3) (8). The most notably affected properties of this tropical Alfisol by soil erosion were total porosity and soil organic matter content. Furthermore, I found maize yield was related significantly to soil properties as follows:

$$Y = 1.79 - 0.007 (E) + 0.70 (O.C.) + 0.07 (M_o) \\ + 0.002 (I_c) \ldots. r = 0.90* \qquad [1]$$

where Y is maize yield in t ha^{-1}, E is accumulative soil loss (t ha^{-1}), O.C. is organic carbon (%), M_o is total porosity (%), and I_c is infiltration capacity (cm). Results from the use of this equation also indicate that adverse effects of erosion on maize yield can be compensated for in part by adding organic matter and by improving total porosity and infiltration capacity. The latter can be achieved by subsoiling.

Geomorphological approach. Researchers often use the geotechniques of evaluating weathering rate and the rate of new soil formation to calculate soil loss tolerance. The latter is defined as the rate of soil erosion that provides for the permanent preservation or improvement of the soil as a resource and is equal to or less than the rate of new soil formation. These techniques, explained by Kirkby (9), are based on "balancing the acceptable rate of soil loss" with the "natural rate of new soil formation." The rate of mechanical erosion is estimated from the weathering rate as follows:

$$T = D\left(\frac{Ps}{1-P_s}\right) \tag{2}$$

where T is the rate of mechanical erosion (mm/year); D is chemical soil formation or solutation weathering (mm/year); and P_s is the degree of weathering, which is 1 for unweathered bedrock. A desirable value for P_s is 0.8. A knowledge of solutational weathering and of P_s values for different parent materials can provide a tentative indication of acceptable soil loss. This technique is more appropriate to evaluate soil loss tolerance than to assess productivity loss due to soil erosion. The geotechnique ignores the edaphological aspects of plant nutrient availability and the importance of organic matter and the clay fraction in plant growth.

Stamey and Smith (20) used a very similar technique to define a usable mathematical function for the computation of tolerable soil loss:

$$T(x,y,t) = T_1 + (T_2 - T_1)/2 + [(T_2 - T_1/2]\text{Cos}\{\pi + [(z - z_1)/(z_2 - z_1)]\pi\} \tag{3}$$

whereT (x,y,t) is tolerable soil loss rate at point (x,y), T_1 and T_2 are lower and upper limits of allowable soil loss rate, T_1 corresponds to soil renewal rate, z_1 and z_2 are minimum allowable and optimum soil depths, and z is the present soil depth.

I used this technique to compute soil loss tolerance for a toposequence in Nigeria (12). Depending upon the effective rooting depth for this particular toposequence in western Nigeria, the soil loss tolerance ranged from as low as 0.05 t ha^{-1} y^{-1} for shallow soils to a maximum of 2 t ha^{-1} y^{-1} for soils with relatively deep, effective rooting depths.

Crop productivity models. Scientists have attempted for many years to establish a relationship between soil properties and soil productivity on the basis of some control variables. The control variables or soil properties that drastically influence soil productivity differ among soils and agroecologies. Control variables considered by various researchers include root growth and water depletion (8, 16, 18). Kiniry and associates (8) described response curves relating sufficiency for root elongation to each soil property. The soil properties considered included potential available water capacity (PAWC), which is influenced by soil texture and structure; bulk density; pH; aeration; and electrical conductivity. On the basis of the sufficiency of these variables in relation to root growth, Kiniry and associates (5) described the productivity index (PI) as follows:

$$PI = \sum_{i=1}^{r} (A \times B \times C \times D \times E \times RI)i \qquad [4]$$

where A is the sufficiency of potential available water storage capacity (SUFFPAW). The researchers assumed that a PAWC of 0.20 or larger was nonlimiting. SUFFPAW was assigned a value of 1 if PAWC $\geq$ 0.20 and PAWC/0.20 if PAWC < 0.20.

B is the sufficiency of aeration (SUFFAER). The air-filled porosity, S, was converted to a form of relative resistance by computing the reciprocal 1/S and then summing relative resistances by the integral $\int_{0}^{z} 1/s\, dz$. Relative conductance, as a reciprocal of relative resistance, is converted to sufficiency classes by assuming the critical value of air-filled porosity to be 0.08.

C is the sufficiency of bulk density (SUFFDB), computed as follows:

SUFFDB $= -0.68\, D_b + 1.88$ for $1.3 < D_b \leq 1.55$
SUFFDB $= -3.32\, D_b + 5.98$ for $1.55 < D_b \leq 1.80$
SUFFDB $= 0$ for $D_b > 1.80$

D is the sufficiency of pH (SUFFPHS), estimated as follows:

SUFFPHS $= 1.0$ if $5.5 < pH \leq 7.5$
SUFFPHS $= 0.16\, pH + 0.12$ if $5.0 < pH \leq 5.5$
SUFFPHS $= 0.446\, pH - 1.31$ if $2.9 < pH \leq 5.0$

E is the sufficiency of electrical conductivity (SUFFEC) in mmhos/cm, expressed as:

SUFFEC = 1.0 if EC ≤ 2.0
SUFFEC = 1.14 − 0.07 EC if 2.0 < EC ≤ 16.0
SUFFEC = 0 if EC ≥ 16.0

RI is predicted root fractions under ideal soil conditions, converted to soil-determined root fractions by multiplying the ideal fraction (RI) for each depth increment by the fractional sufficiency of each soil parameter for that depth increment.

r is the total number of 10-cm-depth increments in the plant-determined rooting depth R.

i is the 10-cm-depth increment number (i = 1,2,3, ... r).

This model has been tested to estimate the relationship between soil loss and crop productivity in western Nigeria (10), Hawaii, ICRISAT (6), and other regions of the tropics and northern latitudes (16). The model is applicable to a wide range of soils in the United States; Pierce and associates (17) used it to evaluate erosion-induced losses in productivity. The model, however, requires appropriate modifications for soils in the tropics. I suggested the following modifications (12):

Sufficiency of organic carbon. The rapid decline in organic matter content of tropical soils causes drastic reductions in soil structure and in plant-available water reserves. Furthermore, the organic matter content is concentrated in the top 10 to 20 cm of soil. Thus, the sufficiency limit of organic matter content needs to be changed.

Presence of skeletal material in the soil profile. Root growth and the water and nutrient retention capacity of the soil is drastically altered by the presence of coarse material exceeding 30 percent by volume (3, 22). Crop response to gravel also depends upon the textural and nutritional properties of the intergravel material.

Rooting depth. The effective rooting depth of most tropical soils rarely exceeds 0.5 m.

It is difficult to use empirical relations if the data base is sketchy. Successful application of these models requires prior knowledge of (a) optimum rooting depth for crops, (b) plant-available soil water reserves, (c) appropriate criteria to assess soil compaction and crop response to different levels of soil compaction, and (d) tolerance to subsoil acidity. El-Swaify and associates arrived at similar conclusions (6). They reported that replacing inapplicable sufficiency curves with simple indexes and adding new parameters, for example, nutrient deficiencies, would improve the application of the PI index to Oxisols in Hawaii.

In addition to PI, a range of processes and economic models have been developed to assess the erosion-productivity relationship. One such model is the erosion-productivity impact calculator, EPIC (19, 20). This model simulates the physical processes involved using routinely available data. Similar to the process models, linear programming models also have been used to evalute erosion-induced loss in productivity (21). I previously reviewed the uses and limitations of such models (13).

Conclusions

It is important to be able to relate crop yields to soil erosion and to the additional inputs needed to maintain the same crop production. The most relevant approach to determine erosion-caused reduction in soil productivity is the direct agronomic approach to assess crop yields on land from which the loss of surface soil has been assessed directly on field runoff plots. The rates of natural erosion can be accelerated by using simulated rainfall. The effects of desurfacing are more drastic for some soils but not as drastic as natural erosion for others. Parametric methods of assessing potential productivity can be useful in providing information relative to the effects of erosion, provided that empirical data relating crop yields to important soil properties exist. If such a data base is not available, the results obtained are of limited use.

REFERENCES

1. Aina, P. O., and E. Egolum. 1980. *The effect of cattle feedlot manure and inorganic fertilizer on the improvement of subsoil productivity.* Soil Science 129: 212-217.
2. American Society of Agricultural Engineers. 1985. *Erosion and soil productivity.* Publication No. 8-85. St. Joseph, Michigan. 287 pp.
3. Babalola, O., and R. Lal. 1977. *Subsoil gravel horizon and maize root growth. I. Gravel concentration and bulk density effects.* Plant Soil 46: 337-346.
4. Babalola, O., and R. Lal. 1977. *Subsoil gravel horizon and maize root growth. II. Effects of gravel size, inter-gravel texture and natural gravel horizon.* Plant Soil 46: 347-357.
5. Dumsday, R. G. 1973. *The economics of some soil conservation practices in the wheat belt of nothern New South Wales and Southern Queensland: A modeling approach.* University of New England Farm Management Bulletin 19:208.
6. El-Swaify, S. A., A. Lo, and F. R. Rijsberman. 1984. *Applicability of the soil*

productivity index to selected soils in Hawaii. In F. R. Rijsberman and M. G. Wolman [editors] *Quantification of the Effect of Erosion on Soil Productivity in an International Context.* Hydraulics Laboratory, Delft, The Netherlands. pp. 95-107.

7. Follett, R. F., and B. A. Stewart. 1985. *Soil erosion and crop productivity.* American Society of Agronomy, Madison, Wisconsin. 533 pp.

8. Kiniry, L. N., C. L. Sirivner, and M. E. Keener. 1983. *A soil productivity index based upon predicted water depletion and root growth.* Research Bulletin 1051. University of Missouri, Columbia. 26 pp.

9. Kirkby, M. J. 1980. *The problem.* In M. J. Kirkby and R.P.C. Morgan [editors] *Soil Erosion.* J. Wiley & Sons, Chichester, England. pp. 1-12.

10. Lal, R. 1976. *Soil erosion on Alfisols in western Nigeria and their control.* Monograph 1. International Institute for Tropical Agriculture, Ibadan, Nigeria. 208 pp.

11. Lal, R. 1981. *Soil erosion problems on Alfisols in western Nigeria. VI. Effects of erosion on experimental plots.* Geoderma 25: 215-230.

12. Lal, R. 1985. *Soil erosion and its relation to productivity in tropical soils.* In S. A. El-Swaify, W. C. Moldenhauer, and A. Lo [editors] *Soil Erosion and Conservation.* Soil Conservation Society of America, Ankeny, Iowa.

13. Lal, R. 1987. *Effects of soil erosion on crop productivity.* CRC Critical review in Plant Sciences 5(4): 303-308.

14. Mbagwu, J.S.C., R. Lal, and T. W. Scott. 1984. *Effects of desurfacing of Alfisols and Ultisols in southern Nigeria. I. Crop performance.* Soil Science Society of America Journal 48: 828-833.

15. Mbagwu, J.S.C., R. Lal, and T. W. Scott. 1984. *Effects of desurfacing of Alfisols and Ultisols in southern Nigeria. II. Changes in soil physical properties.* Soil Science Society of America Journal 48: 834-838.

16. Pierce, F. J., W. E. Larson, and R. H. Dowdy. 1984. *Evaluating soil productivity in relation to soil erosion.* In F. R. Rijsberman and M. G. Wolman *Quantification of the Effect of Erosion on Soil Productivity in an International Context.* Hydraulics Laboratory, Delft, The Netherland. 157 pp.

17. Pierce, F. J., W. E. Larson, R. H. Dowdy, and W.A.P. Graham. 1983. *Productivity of soils: Assessing long-term changes due to erosion.* Journal of Soil and Water Conservation, 38(1): 39-44.

18. Scrivner, C. L., and C. J. Gantzer. 1982. *Soil erosion and soil productivity.* In Proceedings, Conservation Tillage Seminar. University of Missouri, Columbia.

19. Skidmore, E. L. 1979. *Soil loss tolerance.* American Society of Agronomy, Madison, Wisconsin.

20. Stamey, W. L., and R. M. Smith. 1964. *A conservation definition of erosion tolerance.* Soil Science 97: 183-186.

21. Stocking, M. 1984. *Erosion and soil productivity: A review.* Food and Agriculture Organization, Rome, Italy. 102 pp.

22. Vine, P. N., R. Lal, and D. Payne. 1981. *The influence of sands and gravels on root growth of maize seedlings.* Soil Science 131: 124-129.

23. Williams, J. R., and K. G. Renard. 1985. *Assessment of soil erosion and crop productivity with process models (EPIC).* In R. F. Follett and B. A. Stewart [editors] *Soil Erosion and Crop Productivity.* American Society of Agronomy, Madison, Wisconsin.

24. Williams, J. R., K. G. Renard, and P. T. Dyke. 1983. *A new method for assessing erosion's effects on soil productivity.* Journal of Soil and Water Conservation 38:381-383.

10

E. L. Skidmore

Wind erosion

WIND erosion is a serious problem in many parts of the world. Extensive aeolian deposits from past geologic eras also prove it is not a recent phenomenon.

Wind erosion is worse in arid and semiarid areas where the following conditions frequently occur: loose, dry, finely divided soil; a smooth soil surface devoid of vegetative cover; large fields; and strong winds (44). Arid and semiarid lands are extensive. Arid lands comprise about one-third of the world's total land area and are the home of one-sixth of the world's population (37, 50). Areas most susceptible to wind erosion on agricultural land include much of North Africa and the Near East, parts of southern and eastern Asia, the Siberian Plains, Australia, southern South America, and the semiarid and arid portions of North America (44).

Land undergoing desertification becomes vulnerable to wind erosion (85). On pastoral rangeland, composition of pastures subject to excessive grazing during dry periods deteriorates, the proportion of edible perennial plants decreases, and the proportion of annuals increases. The thinning and death of vegetation during dry seasons or droughts increase the extent of bare ground, and surface soil conditions deteriorate, increasing the fraction of erodible aggregates on the soil surface. In rainfed farming areas, removal of the original vegetation and fallow expose the soil to accelerated wind and water erosion.

Extensive soil erosion in the U.S. Great Plains during the last half of the 19th century and in the prairie region of western Canada during the 1920s warned of impending disaster. In the 1930s, a prolonged dry spell culminated in dust storms and soil destruction of disastrous proportions in the prairie regions of both western Canada and the Great Plains (2, 62, 65, 76, 102).

Wind erosion physically removes from the field the most fertile portion of the soil and, therefore, lowers land productivity (35, 68). Some soil from damaged land enters suspension and becomes part of the atmospheric dustload. Hagen and Woodruff (54) estimated that eroding land in the Great Plains contributed 244 million and 77 million tons of dust per year to the atmosphere in the 1950s and 1960s, respectively. Jaenicke (63) estimated the source strength of mineral dust from the Sahara at 260 million tons a year. Dust obscures visibility and pollutes the air, causes automobile accidents, fouls machinery, and imperils animal and human health. Blowing soil also fills road ditches; reduces seedling survival and growth; lowers the marketability of vegetable crops, such as asparagus, green beans, and lettuce; increases the susceptibility of plants to certain types of stress, including diseases; and contributes to transmission of some plant pathogens (33, 58, 59).

Soil erodibility by wind

Scientists recognized early that soil erodibility, the susceptibility or ease of detachment and transport by wind, was a primary variable affecting wind erosion. From wind tunnel tests, Chepil (17) determined relative erodibilities of soils reasonably free from organic residues as a function of apparent specific gravity and proportions of dry soil aggregates in various sizes. Clods larger than 0.84 mm in diameter were nonerodible in the range of windspeeds used in the tests. Since then, the nonerodible soil fraction >0.84 mm, as determined by dry sieving, has been used to indicate erodibility of soil by wind. In an early version of the wind erosion equation (26), the nonerodible soil fraction was one of three major factors developed from results obtained principally with a portable wind tunnel (113, 114, 116).

A dimensionless soil erodibility index, I, was based on the nonerodible fraction, the percentage of clods >0.84 mm in diameter (22, 27). The quantity of soil eroded in wind tunnel tests is governed by the tunnel's length and other characteristics. Therefore, erodibility was expressed on a dimensionless basis so that for a given soil and surface condition the same relative erodibility value would be obtained regardless of wind-tunnel characteristics (24). The soil erodibility index was expressed as follows:

$$I = X_2/X_1 \qquad [1]$$

where X_1 is the quantity eroded from soil containing 60 percent of clods > 0.84 mm and X_2 is the quantity eroded under the same set of conditions from soil containing any other proportion of clods > 0.84 mm. The soil erodibility index, I, gave a relative measure of erodibility, but actual soil loss by wind was not known.

Therefore, during the severe wind erosion seasons of 1954-1956, from January through April, Chepil studied 69 fields in western Kansas and eastern Colorado to determine the quantity of soil loss for any field erodibility as determined from various field conditions (24). The average depth of soil eroded usually was indicated by the depth to which crowns and roots of plants were exposed.

Seasonal loss was converted to annual soil loss, and relative field erodibility for each field was determined by procedures previously outlined (23, 26, 27). The relation between annual soil loss and relative field erodibility was as follows:

$$Y = aX^b - 1/cd^x \qquad [2]$$

where Y is annual soil loss (tons/acre); X is the dimensionless relative field erodibility; and a, b, c, and d are constants equal to 140, 0.287, 0.01525, and 1.065, respectively. Chepil (24) recognized that inaccuracies in measuring relatively small annual soil losses from depth of soil removal made converting relative field erodibility to annual soil loss by equation 2 highly approximate.

When a field is smooth, bare, wide, unsheltered, and noncrusted, its relative erodibility is equal to the erodibility index defined by equation 1. To obtain potential annual soil loss in tons per acre, I is substituted for x in equation 2. Equation 2 was multiplied by one-third, then used to generate a table (109) for erodibility of soils with different percentages of nonerodible fractions > 0.84 mm (Table 1).

A more reliable and technically sound procedure is needed to estimate or predict the erodibility index without making physical measurements. This would save time and expense and provide a means to estimate erodibility more accurately.

In current practice, scientists often estimate soil erodibility by grouping soils, mostly according to predominant soil textural class (Table 2).

Because of the utility of predicting soil erodibility from easily obtainable soil properties, table 2, or a similar one, has been used extensively. Problems associated with using table 2 to estimate soil erodibility include the transience of dry soil aggregates > 0.84 mm within a given wind erodibility group (WEG).

Actual erodibility is extremely dynamic and varies seasonally, yearly, and as the result of management operations. In a study on the effects of season on soil erodibility, Chepil (18) found erodibility always was higher in the spring than in the previous fall if the soil had received moisture occasionally during the winter. But the increases were not of the same magnitude for all soils. The greatest increase in erodibility from fall to spring occurred in the finest textured soil, the least in the coarsest. Sandy loam was highly erodible in both fall and spring. Clay was least erodible in the fall, but about as highly erodible as sandy loam in the spring. The intermediate-textured soils had an intermediate erodibility in both spring and fall.

Grouping is discontinuous and results in large, discrete jumps in erodibility with textural change. For example, the soil erodibility indexes for loamy very fine sand (WEG 2) and very fine sandy loam (WEG 3) are 300 and 193 Mg/ha/yr, respectively. All other soils classed in those wind erosion groups have the same erodibility indexes. Do they have the same erodibility? It would be better to predict percentage of aggregates > 0.84 mm and then use table 1. A procedure must be devised for realistically predicting dry soil-

Table 1. Soil erodibility, I, for soils with different percentages of nonerodible fractions as determined by standard dry sieving (109).

Percent	Soil Erodibility by Percentage of Dry Soil Fractions > 0.84 mm									
	0	1	2	3	4	5	6	7	8	9
	Mg/ha									
0	-	695	560	493	437	404	381	359	336	314
10	300	294	287	280	271	262	253	244	238	228
20	220	213	206	202	197	193	186	182	177	170
30	166	161	159	155	150	146	141	139	134	130
40	126	121	117	114	112	108	105	101	96	92
50	85	80	75	70	65	61	28	54	52	49
60	47	45	43	40	38	36	36	34	31	29
70	27	25	22	18	16	13	9	7	7	4
80	4	-	-	-	-	-	-	-	-	-

aggregate status that accounts for yearly and seasonal fluctuations and dominant soil properties influencing erodibility.

The aggregate status of the soil at any instant in time is the result of many aggregate-forming and degrading processes. These processes comprise a complex interrelationship of physical, chemical, and biological reactions. Aggregation may be the breakdown of clods into more favorable size, or it may be the formation of aggregates from finer materials.

Another factor to be considered in assessing or predicting the aggre-

Table 2. Descriptions of wind erodibility groups (105).

WEG	Predominant Soil Texture Class of Surface Layer	Dry Soil Aggregates > 0.84 mm (%)	Wind Erodibility Index, I (Mg/ha)
1	Very fine sand, fine sand, or coarse sand	1	695
		2	560
		3	493
		5	404
		7	359
2	Loamy very fine sand, loamy fine sand, loamy sand, loamy coarse sand, or sapric soil materials	10	300
3	Very fine sand loam, fine sandy loam, sandy loam, or coarse sandy loam	25	193
4	Clay, silty clay, noncalcareous clay loam, or silty clay loam with more than 35 percent clay content	25	193
4L	Calcareous loam and silt loam or calcareous clay loam and silty clay loam	25	193
5	Noncalcareous loam and silt loam with less than 20 percent clay content or sandy clay loam, sandy clay, and hemic organic soil materials	40	126
6	Noncalcerous loam and silt loam with more than 20 percent clay content or noncalcereous clay loam with less than 35 percent clay content	45	108
7	Silt, noncalcareous silty clay loam with less than 35 percent clay content, and fibric organic soil material	50	85
8	Soils not susceptible to wind	> 80	0

gate status or erodibility of a soil is the influence of cropping history and tillage. Page and Willard (82) found that the degree of aggregation in a corn/oats/alfalfa-bromegrass/alfalfa-bromegrass rotation is two to three times greater than that for continuous corn. Cropping systems that included continuous small grain, continous row crops, and rotations including fallow showed no significant differences in water-stable aggregation (81). Soils broken out of native sod lost much of their aggregation in the surface-tilled zone (81, 92, 100). Skidmore and associates, in a study of soil physical properties as influenced by management of residues from winter wheat and grain sorghum, found that grain sorghum or wheat management treatments did not influence most of the soil physical properties measured (92). However, the aggregate status differed among crops. Soil aggregates from sorghum plots were smaller, more fragile, less dense, and more wind-erodible than aggregates from wheat plots. Harris and associates (57) reported that agronomic systems affect aggregation significantly but that interpreting controlling mechanisms is complicated by the diversity of factors through which the effects are manifest.

Inability to predict both aggregate status and the weather undoubtedly influenced Woodruff and Siddoway's definition of soil erodibility (109): "the potential average annual soil loss from a wide, unsheltered, isolated field...for the climate in the vicinity of Garden City, Kansas." In spite of temporal variation of soil aggregate status, Woodruff and Siddoway suggested that soil erodibility can be estimated by standard dry sieving and use of table 1. Use of sieving results assumes that the values determined (percent > 0.84 mm) "characterize" a soil during the critical erosion period for the time domain of the wind erosion equation (109).

For determining percentages of dry soil fractions > 0.84 mm, Chepil and Woodruff (27) recommended the rotary sieve. A conventional and more readily available flat sieve may be used, but results with it are less accurate than with a rotating sieve.

Researchers should use the following procedure when using a flat sieve:

▶ Obtain 1 kg samples from the 0- to 2-cm surface layer when soil is reasonably dry. If soil is not near air dryness, dry it in the laboratory before sieving.

▶ Weigh the sample and sieve it on a 0.84-mm (No. 20), 20.3-cm (8-inch) diameter sieve until the aggregates < 0.84 mm diameter have passed through the sieve. Be careful not to fragment aggregates during sieving. Weigh the amount of sample remaining on the sieve.

▶ Calculate the mass fraction of the total sample that was retained on the sieve and use table 1 to determine soil erodibility.

Suppose from replicated sievings from a sample site that the total amount of air-dried soil for each sieving was 1,035, 945, 850, and 990 grams and the respective amounts retained on the 0.84 mm sieve after sieving were 370, 227, 200, and 250 grams. Therefore, percentages of dry soil fractions > 0.84 mm would be 26.1, 24.0, 20.7, and 25.3, respectively. Corresponding soil erodibility values from table 1 would be 186, 197, 213, and 193 Mg/ha, respectively; the mean would be 197 Mg/ha.

Wind erosivity

Chepil and associates (25) proposed a climatic factor to determine average annual soil loss for climatic conditions other than those occurring when the relationship between wind-tunnel erosion and field erosion was obtained. It is an index of wind erosion as influenced by moisture content in surface soil particles and average windspeed. The windspeed term of the climatic factor was based on the rate of soil movement being proportional to average windspeed cubed (8, 15, 115). The soil moisture term was developed on the basis that the erodibility of soil varies inversely with the square of the equivalent water content in the near-surface soil, which was assumed to vary as the Thornthwaite index (20).

The climatic factor was expressed as follows:

$$C = 386 \frac{\bar{u}^3}{(PE)^2} \quad [3]$$

where $\bar{u}$ is the mean annual windspeed corrected to 9.1 m and PE is the Thornthwaite (103) index. The 386 value indexes the factor to conditions at Garden City, Kansas.

Thornthwaite developed the climatic index to evaluate precipitation effectiveness. An equation was fitted to rather limited data that expressed the P/E ratio to temperature and precipitation as follows:

$$P/E = 0.316 \left(\frac{P}{1.8T + 22}\right)^{10/9} \quad [4]$$

where P is the mean monthly precipitation in mm, E is the monthly evaporation in mm, and T is temperature in C°. Monthly values were added to obtain an annual value, which was multiplied by 10 to give:

$$\text{PE index} = 3.16 \sum_{i=1}^{12} \left(\frac{P_i}{1.8T_i + 22} \right)^{10/9} \qquad [5]$$

Equation 5 was evaluated and used in equation 3 to determine climatic factors for wind erosion at many locations in the United States (25, 69, 98).

As the PE index gets smaller as precipitation declines, as in arid regions, the climatic factor in equation 3 approaches infinity. In application, an upper limit is set by restricting minimum monthly precipitation to 13 mm (69). Monthly climatic factors also were calculated using an annual PE index with monthly mean windspeed (108).

The Food and Agriculture Organization approached the problem of the climatic factor becoming a large value in arid conditions differently (45). Agency researchers modified the Chepil and associates' index (25) as follows:

$$C^1 = 1/100 \sum_{i=1}^{12} \bar{u}^3 \left(\frac{ETP_i - P_i}{ETP_i} \right) d \qquad [6]$$

where $\bar{u}$ is the mean monthly windspeed at a 2-m height, ETP is potential evapotranspiration, P is precipitation, and d is the total number of days in the month. In this case, as precipitation approaches zero, windspeed dominates the climatic factor. Conversely, as precipitation approaches ETP, the climatic factor approaches zero. The influence of soil water in the FAO version is less than the squared influence of soil water demonstrated by Chepil (20).

I handled the influence of soil water differently and included a windspeed probability density function as follows (90):

$$CE = \varrho \int_{R}^{\infty} \left[u^2 - R^2 \right]^{3/2} f(u)du \qquad [7]$$

where,

$$R = u_t^2 + \gamma^1/\varrho a^2 \qquad [8]$$

and CE is the wind erosion climatic erosivity, which is directly proportional to mass flow rate of an all erodible material; ϱ is air density; u and u_t are windspeed and threshold windspeed, respectively;

γ^1 is the cohesive resistance of absorbed water; and a is a combination of constants, $k/\ln(z/z_0)$, for $k = 0.41$, $z = 10$ m, $z_0 = 0.05$ m; thus, $a = 0.0774$.

The value for γ is approximated as follows:

$$\gamma = 0.5 \, \psi^2 \tag{9}$$

where ψ is the equivalent soil water content, fraction of water (by mass or volume) in the soil, divided by fraction of water in the same soil at $-1,500$ J/kg (20, 90). It was assumed that equivalent surface water content was approximated by the ratio of precipitation to potential evaporation. The ratio of precipitation to evaporation can be approximated by the Thornthwaite PE index or the inverse of the dryness ratio (12, 56). The dryness ratio, D, is defined as follows:

$$D = Rn/(LP) \tag{10}$$

where Rn is net radiation, L is latent heat of evaporation, and P is precipitation. The dryness ratio at a given site indicates the number of times the net radiative energy could evaporate the precipitation over the same time interval.

The windspeed probability density function, equation 7, can be expressed as a Weibull distribution:

$$f(u) = (k/c) \, (u/c)^{\,k\text{-}1} \exp[-(u/c)^k] \tag{11}$$

where c and k are scale and shape parameters, respectively. Parameter c has units of velocity and k is dimensionless (3, 66). Weibull parameters have been determined from windspeed distribution summaries at many locations in the U.S. Great Plains (53).

Equation 7, with f(u) defined by equation 10, can be integrated straightforwardly when $k = 2$ as follows:

$$CE = 1.33 \varrho c^3 \exp\left[-R/c^2\right] \tag{12}$$

where R is defined by equation 8.

The summation procedure for evaluating equation 7 can be written as follows:

$$CE = \varrho \sum_{\substack{u_i^2 + 0.5 > R}}^{n} \left(u_{i+0.5}^2 - R\right)^{3/2} \left[F\left(u_{i+1}\right) - F\left(u_i\right)\right] \tag{13}$$

where $F(u_i)$ is the cumulative distribution function:

$$F(u_i) = 1 - \exp\left[-(u_i/c)^k\right] \tag{14}$$

When mean windspeed is available but the data from which the mean was calculated are not, the Weibull parameters can be estimated.

Studies have shown that the Weibull scale parameter was about 12 percent larger than mean windspeed and the Weibull shape parameter was a function of the scale parameter (64, 90). Thus, if only mean windspeed is known, a reasonable estimate of Weibull distribution can be obtained as follows:

$$c = 1.12 \ \bar{u} \qquad\qquad [15]$$

and

$$k = 0.52 + 0.23c \qquad\qquad [16]$$

Equations 7 and 13 express wind power, W m^{-2}. When multiplied by the time duration in the accounting period represented by f(u), they give erosive wind energy. This is the energy of the wind in excess of that necessary to overcome threshold shear stresses represented by R. Erosive wind energy is a useful parameter to evaluate the climatic factor for the wind erosion equation.

Suppose one wishes to know an appropriate climatic factor for a 30-day period of given conditions: mean windspeed = 5.8 m s^{-1}, average precipitation = 80 mm, net radiation = 490 MJ m^{-2}. Then, from equations 15 and 16, c and k are estimated to be 6.5 m s^{-1} and 2.0, respectively. The dryness ratio calculated from equation 10 is 2.5 for heat of vaporization of 2.45 MJ kg^{-1}.

Thus,

$$R = u_t^2 + \gamma/\varrho a^2 = 36 + 11 = 47 \ m^2 \ s^{-2} \qquad\qquad [17]$$

Equation 13 could be used to calculate CE. However, because k = 2.0, equation 12 was used to calculate CE as follows:

$$CE = 1.33 \ \varrho c^3 \exp\left[-(R/c^2)\right] = 144 \ W \ m^{-2} \qquad\qquad [18]$$

Therefore, the erosive wind energy for the 30-day period would be as follows:

$$CE \times time = 144 \ W \ m^{-2} \times 8.64 + 10^4 \ sd^{-1} \times 30d = 373 \ MJ \ m^{-2} \qquad [19]$$

If the conditions given in this example 30-day period were to prevail for an entire year, then the erosive wind energy would be 4,538 MJ m^{-2}. That wind energy, compared to the reference of 8,100 MJ m^{-2}, gives a climatic factor of 56. Also, from figure 1, for

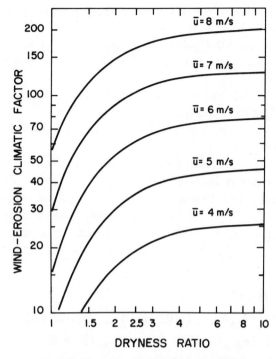

Figure 1. Wind erosion climatic factor as influenced
by dryness ratio and mean windspeed (90).

a dryness ratio of 2.5 and a mean windspeed of 5.8 m s^{-1}, the
climatic factor is 56.

Ridge roughness

Chepil and Milne (32) investigated the influence of surface
roughness on intensity of drifting dune materials and cultivated soils.
They found that the initial intensity of drifting was always much
less over a ridged surface. Ridging cultivated soil reduced the severity
of drifting, but ridging highly erodible dune material was less effective because ridges disappeared rapidly. The rate of flow varied inversely with surface roughness. Armbrust and associates (7) studied
the effects of ridge-roughness equivalent on the total quantity of
eroded material from soils exposed to different friction velocities.
From their data, a curve can be constructed showing the relation-

tween the relative quantity of eroded material and the ridge-roughness equivalent. Presumably, this was the origin of the chart (*109, figure 4*) showing a soil ridge-roughness factor as a function of soil ridge roughness, so that ridging may reduce wind erosion up to 50 percent.

Ridge roughness estimates the fractional reduction of erosion caused by ridges of nonerodible aggregates. It is influenced by ridge spacing and ridge height and is defined relative to a 1:4 ridge-height-to-ridge-spacing ratio.

Tables were prepared of ridge-roughness factors for various combinations of ridge heights and spacings (*88*). Hayes (*60*) suggested evaluating fields as either smooth, semiridged, or ridged and then assigning 1.0, 0.75, and 0.50, respectively, as soil ridge-roughness factors. Williams and associates (*106*) fitted equations to the curve of Woodruff and Siddoway (*109*) to express the ridge-roughness factor as follows:

$$K = 1.0, \; HR^2/IR < 0.57 \tag{20}$$

$$K = 0.913 - 0.153 \ln (HR^2/IR), \; 0.57 < (HR^2/IR) < 22.3 \tag{21}$$

$$K = 0.336 \exp (.013 \; HR^2/IR), \; (HR^2/IR) \geq 22.3 \tag{22}$$

where HR and IR are ridge height and ridge spacing, respectively, in mm. A field with ridges 100 mm high and spaced 400 mm apart has $HR^2/IR = 25$. Because $25 > 22.3$ and using equation 22, the ridge-roughness factor $K = 0.5$.

Field length

Chepil and Milne (*32*) reported that the rate of soil movement started at zero on the windward side of fields or field strips and increased with distance downwind. Later, Chepil (*16*) found that the cumulative rate of soil movement with distance away from the windward edge of eroding fields was the main cause of steadily increasing amounts of erodible particles, increasing abrasion, and gradual reduction in the rate of soil flow with distance downwind "avalanching."

Rate of soil flow increased with distance downwind across an eroding field. If the field were large enough, soil flow reached the maximum that a wind of a given velocity could carry. Beyond that point, the rate of flow remained essentially constant (*21*). That maximum was about the same for soil of any texture—about 50 $gm^{-1}s^{-1}$ for a 17 m s^{-1} wind at 10 m. The rate of increase for various soil textures was the same as the order of erodibility for soil texture classes.

The distance required for soil flow to reach the maximum that a wind of a given velocity can carry varies inversely with the erodibility of a field surface. The more erodible the surface, the shorter the distance to reach maximum flow (23).

Chepil (23) related relative wind erodibility to the distance required for soil flow to reach a maximum. In his earlier work (16, 21, 32), he presented data for the rate of soil movement as a function of distance from the windward edge of the field for soils that varied widely in erodibility. He converted the relative surface erodibility, based on four factors—soil cloddiness, crop residue, ridge-roughness equivalent, and soil erodibility—to relative field erodibility, based on additional factors—wind barrier, width of field, and wind direction (23). These functional relationships between field erodibility and field width with the many associated factors gave rise to how the field length term was used in the wind erosion equation (28, 109).

Originally, field length was considered as the distance across a field in the prevailing wind erosion direction (109). However, sometimes almost as much wind occurs from one direction as from another, so there is essentially no prevailing wind erosion direction. In these cases, researchers used the preponderance of wind erosion forces in the prevailing wind erosion direction to assess equivalent field length (87, 98). Later, from a more detailed analysis, tables were prepared that give wind erosion direction factors, numbers that when multiplied by field width give median travel distance as a function of preponderance of wind erosion forces in the prevailing direction and deviation of prevailing wind erosion direction from perpendicular to direction of field length (90).

In some of the modeling efforts, the procedure for determining L for use in the wind erosion equation was simplified by ignoring wind direction distributions. Cole and associates (34) suggested the following:

$$L = \begin{cases} w \sec \theta \\ l \csc \theta \end{cases} \qquad \begin{matrix} L \leq (l^2 + w^2)^{\frac{1}{2}} \\ \text{otherwise} \end{matrix} \qquad [23]$$

where w and l are the small and large dimensions, respectively, of a rectangular field and θ is the angle between side w and the prevailing wind erosion direction. As θ varies through $\pi/2$ radians, L will range from w to l, with a maximum equal to the main diagonal of the field. The procedure Williams and associates (106) used in EPIC, the erosion-productivity impact calculator, was as follows:

$$L = \frac{lw}{l|\cos(\pi/2 + \alpha - \phi)| + w|\sin(\pi/2 + \alpha - \phi)|} \quad [24]$$

where l and w are the large (length) and small (width) dimensions, respectively, of a rectangular field, α is the wind direction clockwise from north in radians, and ϕ is the clockwise angle between field length and north in radians. Using equation 24, L = 236 m for a rectangular field where l = 1,000 m, w = 200 m, $\alpha = \pi/4$ radians, and $\phi = 0$.

Vegetative factor

Scientists realized early the value of crop residue for controlling wind erosion and reported quantitative relationships (14). From wind tunnel tests on plots especially prepared to obtain a range of vegetative cover and soil structure, Englehorn and associates (39) found the expotential relationship that best expressed their results. Subsequent studies (19, 26, 27) expressed the relationship in the form $x = aI/(RK)^b$, where x is the wind tunnel erodibility; I is the soil erodibility index (percent of clods > 0.84 mm); R is the dry weight of crop residue in pounds/acre; K is the ridge-roughness equivalent; and a and b are constants.

Amounts of wheat straw needed to protect most erodible dune sands and less erodible soils against strong winds were established (31). Standing stubble was much more effective than flattened stubble (29). Standing sorghum stubble with rows perpendicular to wind direction controlled wind erosion more effectively than rows parallel to wind direction (39, 97).

Siddoway and associates (86) quantified the specific properties of vegetative covers influencing soil erodibility and developed regression equations relating soil loss by wind to selected amounts, kinds, and orientation of vegetative covers, wind velocity, and soil cloddiness. They found a complex relationship among the relative effectiveness of different kinds and orientation of residue. The relative value of kinds and orientations of residue in controlling erosion must be qualified by soil, wind velocity, and variable characteristics of the residues. Generally, Siddoway and associates concluded that (a) on a weight basis, fine-textured residues were more effective than coarse-textured residues; (b) any orientation of residue, except flattened residue, reduced wind erosion; and (c) fine-leafed crops, such as grasses and cereals, provided a high degree of erosion control per unit weight.

Those studies led to the relationship developed by Woodruff and Siddoway (109) showing the influence of an equivalent vegetative cover of small grain and sorghum stubble for various orientations (flat, standing) and heights, then relating soil loss to equivalent vegetative cover.

Efforts to evaluate the protective role of additional crops have continued. In wind tunnel tests, Lyles and Allison (70, 71) determined equivalent wind erosion protection provided by selected range grasses and crop residues. They found high simple correlation coefficients from an equation of the form:

$$(SG)_e = aX^b \qquad\qquad [25]$$

where $(SG)_e$ is the flat small-grain equivalent, X is the quantity of residue or grass to be converted, and a and b are constants. Tables 3 and 4 show prediction equation coefficients.

It is not practical in testing all combinations of crops and residues to determine their protection value as flat small-grain equivalents. Therefore, a practice is needed to estimate the protection values of crops and residues not tested. Hayes (59) suggested that if any residue is not represented researchers should use a curve for the crop most like the crop in question.

Lyles and Allison (71) correlated measurable parameters, which describe crop residues, in several combinations to obtain an equation for predicting the flat small-grain equivalent of flat, random residues as follows:

$$(SG)_e = 0.162 \ Rw/d + 8.708 \ (Rw/d\gamma)^{1/2} - 271$$
$$r^2 = .92 \qquad\qquad [26]$$

where $(SG)_e$ is the flat small-grain equivalent (kg/ha), Rw is the residue amount to be converted (kg/ha), d is the average stalk diameter (cm), and γ is the average specific weight of the stalk (g/cm^3). Winter wheat, rape, soybeans, cotton, and sunflowers were used in developing equation 26.

Table 3. Coefficients in prediction equation $(SG)_e = aR_w^b$ for conversion of crop residues to an equivalent quantity of flat small-grain residue, both in kg/ha (71).

Crop Residue	Surface Orientation	Height (cm)	Length (cm)	Row Spacing (cm)	Row Orientation To Flow	Prediction a	Equation b	Coefficients r^2
Winter wheat	Standing	25.4	-	25.4	Normal	4.306	0.970	0.997
Rape	Standing	25.4	-	25.4	Normal	0.103	1.400	0.990
Cotton	Standing	34.3	-	76.2	Normal	0.188	1.145	0.998
Sunflowers	Standing	43.2	-	76.2	Normal	0.021	1.342	0.994
Winter wheat	Flat-random	-	25.4	-	-	7.279	0.782	0.993
Soybeans	Flat-random	-	25.4	-	-	0.167	1.173	0.993
Rape	Flat-random	-	25.4	-	-	0.064	1.294	0.997
Cotton	Flat-random	-	25.4	-	-	0.077	1.168	0.998
Sunflowers	Flat-random	-	43.2	-	-	0.011	1.368	0.993
Forage sorghum	Standing	15.9	-	76.2	Normal	0.353	1.124	0.995
Silage corn	Standing	15.9	-	76.2	Normal	0.229	1.135	0.998
Soybeans	1/10 standing	6.4	-	76.2	Normal	0.016	1.553	0.991
	9/10 flat-random	-	25.4	-	-			

Table 4. Coefficients in prediction equation $(SG)_e = aX^b$ for conversion of range grasses to an equivalent quantity of flat small-grain residue, both in kg/ha (70).

Grass Species	Grazing Management	Grass Height (cm)	Prediction a	Equation b	Coefficients r^2
Blue grama	Ungrazed	33.0	0.60	1.39	0.98
Buffalograss	Ungrazed	10.2	1.40	1.44	0.97
Big bluestem	Properly grazed	15.2	0.22	1.34	0.99
Blue grama	Properly grazed	5.1	1.60	1.08	0.99
Buffalograss	Properly grazed	5.1	3.08	1.18	0.99
Little bluestem	Properly grazed	10.2	0.19	1.37	0.99
Switchgrass	Properly grazed	15.2	0.47	1.40	0.99
Western wheatgrass	Properly grazed	10.2	1.54	1.17	0.99
Big bluestem	Overgrazed	2.5	4.12	0.92	0.99
Blue grama	Overgrazed	2.5	3.06	1.14	0.99
Buffalograss	Overgrazed	1.5	2.45	1.40	0.99
Little bluegrass	Overgrazed	2.9	0.52	1.26	0.99
Switchgrass	Overgrazed	2.5	1.80	1.12	0.99
Western wheatgrass	Overgrazed	2.5	3.93	1.07	0.99

Until recently, all small-grain equivalence data have been limited to dead crop residue or dormant grass. Armbrust and Lyles (6) reported flat small-grain equivalents for five growing crops—corn, cotton, grain sorghum, peanuts, and soybeans, as follows:

$$(SG)_e = a_1 Rw^{b_1} \tag{27}$$

where $(SG)_e$ is the flat small-grain equivalent and Rw is the above-ground dry weight of the crop to be converted, both in kg/ha, and a_1 and b_1 are constant coefficients for each crop. They found that if only rough estimates of $(SG)_e$ are needed, an average coefficient could be used. An average equation determined from pooling all crop data with rows running perpendicular to wind direction gave 8.9 and 0.9, respectively, for a_1 and b_1.

Suppose one wishes to know the equivalent flat small-grain residue for a field with grain sorghum growing in 400 kg/ha of flat, random winter wheat residue when the dry weight of the growing grain sorghum is 83 kg/ha and the grain sorghum is growing in rows perpendicular to the expected wind. Therefore, $(SG)_e$ for the growing sorghum, from equation 27, would be as follows:

$$(SG)_e = 8.9(83)^{.9} = 475 \tag{28}$$

and, from table 3, $(SG)_e$ for the wheat residue would be

$$(SG)_e = 7.3(400)^{.8} = 880. \tag{29}$$

However, because of nonlinear relationships, the flat small-grain equivalents are not strictly additive. When more than one crop contributes to the residue, it is better to combine the calculation into a single equation as follows:

$$(SG)_e = a_1{}^{p_1} a_2{}^{p_2} (Rwt)^{b_1 p_1 + b_2 p_2} \tag{30}$$

where p_1 and p_2 are fractions of total residue, Rwt, and a_1, a_2 and b_1, b_2 are constant coefficients for respective crops as in equation 27. For our example as follows:

$$(SG)_e = (8.9)^{.172} (7.3)^{.828} (483)^{(.9)(.172) + (.8)(.828)}$$
$$= 1{,}190 \text{ kg/ha} \tag{31}$$

Either the equivalent flat small-grain or vegetative factor is needed for the various procedures to estimate wind erosion. The relationship between equivalent flat small-grain and vegetative cover was given graphically by Woodruff and Siddoway (109). Williams and associates (*106*) fitted an equation to the graphical relationship as follows:

$$V = 0.2533 \; (SG)_e^{1.363} \tag{32}$$

Therefore,

$$V = 0.2533 \; (1{,}180)^{1.363} = 3{,}896 \; Mg \; ha^{-1} \tag{33}$$

A wind erosion model

Researchers and scientists have used a wind erosion equation proposed by Woodruff and Siddoway (*109*), with various modifications, for the past 20 years. The model was developed as a result of investigations to understand the mechanics of the wind erosion process, to identify major factors influencing wind erosion, and to develop wind erosion control methods. The general functional relationship between the independent variable E, the potential average annual soil loss, and the equivalent variables or major factors is as follows:

$$E = f(I, K, C, L, V) \tag{34}$$

where I is the soil erodibility index, K is the soil ridge-roughness factor, C is the climatic factor, L is the unsheltered median travel distance of wind across a field, and V is the equivalent vegetative cover. These factors were discussed in more detail earlier.

Solving the functional relationships of the wind erosion equation as presented by Woodruff and Siddoway (*109*) required the use of tables and figures. The awkwardness of the manual solution prompted a computer solution (*43, 99*) and development of a slide-rule calculator (*89*).

The model has been adapted for use with personal computers (*55*) and interactive programs (*40*). Cole and associates (*34*) adapted the Woodruff and Siddoway (*109*) model for simulating daily soil loss by wind erosion as a submodel in EPIC (*106*). The latter version was simplified by fitting equations to the figures of Woodruff and Siddoway (*109*).

Solution of the wind erosion equation gives the expected amount of erosion from a given agricultural field. A second application of the equation is to specify the amount of erosion that can be tolerated and then solve the equation to determine the conditions required to limit soil loss to the specified amount, for example, the amount of residue, field width, etc. Conservationists have used the equation widely for both of these applications.

U.S. Soil Conservation Service field workers have used the equation extensively to plan wind erosion control practices (59). Hayes (58) also used the wind erosion equation to estimate crop tolerance to wind erosion. The equation is a useful guide to wind erosion control principles as well (13, 80, 111). Other uses of the equation include (a) determining spacing for barriers in narrow strip-barrier systems (52), (b) estimating fugitive dust emissions from agricultural and subdivision lands (83, 107), (c) predicting horizontal soil fluxes to compare with vertical aerosol fluxes (49), (d) estimating the effects of wind erosion on soil productivity (67, 106), (e) delineating those croplands in the Great Plains where various amounts of crop residues may be removed without exposing the soil to excessive wind erosion (96), and (f) estimating erosion hazards in a national inventory (104).

The following example of how to use the wind erosion equation to predict expected soil loss employs the variables used earlier, that is, $I = 197$ t ha^{-1}yr^{-1}; $K = 0.5$, $C = 56$, $L = 236$ m, and $V = 3.9$ Mg ha^{-1}. To determine the erosion estimate, however, requires a special combination of the factors. Several approaches are possible to find the solution: graphs, figures, tables, slide rule, or computer. Here, I use the procedure presented by Williams and associates (106). This procedure is done stepwise, but it has been simplified computationally by fitting equations to figures of Woodruff and Siddoway (109). The first step (E1) is to determine soil erodibility, I. Steps E2 and E3 are determined by multiplying the factors indicated as follows:

$$E2 = IK = 197 \times 0.5 = 93 \text{ Mg ha}^{-1} \text{ yr}^{-1} \qquad [35]$$
$$E3 = IKC = 93 \times .56 = 52 \text{ Mg ha}^{1} \text{ yr}^{-1} \qquad [36]$$

E4, the inclusion of field length, is

$$E4 = (WF^{0.348} + E3^{0.348} - E2^{0.348})^{2.87} = 33 \text{ Mg ha}^{-1} \text{ yr}^{-1} \qquad [37]$$

where

$$WF = E2(1.0 - 0.122(L/Lo)^{-0.383} \exp(-3.33 \text{ } L/Lo) = 64 \qquad [38]$$

and

$$Lo = 1.56 \times 10^{6}(E2)^{-1.26} \exp(-0.00156 \text{ } E2) = 4,465 \text{ m} \qquad [39]$$

WF is a field length factor; it accounts for the influence of field length on reducing the erosion estimate. Lo is the maximum field length for reducing the wind erosion estimate.

Parameters Ψ_1 and Ψ_2 are functions of the vegetative cover factor described by the equations:

$$\begin{aligned}\Psi_1 &= \exp(-0.759V - 4.74 \times 10^{-2}V^2 + 2.95 \times 10^{-4}V^3) \\ &= 0.026\end{aligned} \quad [40]$$

$$\begin{aligned}\Psi_2 &= 1 + 8.93 \times 10^{-2}V + 8.51 \times 10^{-3}V^2 - 1.5 \times 10^{-5}V^3 \\ &= 1.469\end{aligned} \quad [41]$$

where V is in Mg ha^{-1} and for our example, from equation 33, has the value of 3.9 Mg ha^{-1}. Therefore,

$$E5 = \Psi_1 E4^{\Psi_2} = 0.026 (33)^{1.469} = 4.4 \text{ Mg ha}^{-1} \text{ yr}^{-1} \quad [42]$$

The estimate of 4.4 Mg ha^{-1} yr^{-1} given by equation 42 is the annual rate of expected erosion during the 30-day period represented by the climatic factor C. To determine the expected erosion during the accounting period, it is necessary to multiple the given estimate by the fraction of the average annual total erosive wind energy occurring during the 30-day accounting period.

Management effects

Rough, cloddy surface. Tillage operations that leave furrows or ridges reduce wind erosion, as discussed earlier. When ridges are nearly gone, vegetative cover is depleted, and the threat of wind erosion continues, a rough, cloddy surface resistant to the force of wind can be created on many cohesive soils with appropriate "emergency tillage." For example, Lyles and Tatarko (73) found that chiseling of growing winter wheat on a silty clay soil greatly increased nonerodible surface aggregates without influencing grain yields. Farmers can use listers, chisels, cultivators, one-ways with two or three disks removed at intervals, and pitting machines to bring compact clods to the surface. Emergency tillage is most effective when done at right angles to the prevailing wind direction. Because clods eventually disintegrate, sometimes rapidly, emergency tillage offers only temporary wind erosion control at best (111, 112).

Residue. Living vegetation or residue from harvested crops protects the soil against wind erosion. Standing crop residues provide nonerodible elements that absorb much of the shear stress in the boundary layer. When vegetation and crop residues are sufficiently high and dense to prevent intervening soil-surface drag from exceeding threshold drag, soil will not erode. Rows perpendicular to wind direction control wind erosion more effectively than do rows parallel to wind direction (39, 97). Flattened stubble, though not as effective as standing stubble, also protects the soil from wind erosion (29).

Soon after the disastrous "dirty thirties" in the U.S. Great Plains, researchers demonstrated that stubble-mulching was a feasible method of reducing wind erosion on cultivated land (38). Stubble-mulching is a crop residue management system using tillage, generally without soil inversion, usually with blades or V-shaped sweeps (77, 78).

Other reduced and modified tillage systems have evolved with efforts to maintain residue on the soil surface. Chemical fallow (11) and ecofallow (41) systems use herbicides or herbicides and subsurface tillage during fallow periods to conserve a large quantity of residue on the surface.

Directly seeding small grains and other crops into stubble without a fallow period and without tillage is being studied and shows promise. The advantages of this system, compared with the tillage systems designed to preserve residues on the surface, include the following: (a) the standing stubble is needed for erosion control until the seeded crop produces enough cover to control erosion; (b) standing stubble more effectively controls erosion than does an equal quantity of flattened residue; (c) standing stubble, because it is not in direct contact with the soil, is less subject to decomposition than is stubble that has been tilled and mixed with the soil; and (d) without tillage, the soil is not pulverized.

The goal is to leave a desirable quantity of plant residue on the soil surface at all times. Residue is needed for a period of time after the crop is planted to protect the soil from erosion and improve water infiltration. The residue used is generally that remaining from a previous crop. Efforts continue to evaluate the residue needed to control wind erosion (75, 91, 96).

Stabilizers. Researchers have evaluated various soil stabilizers to find suitable materials and methods to control wind erosion (4, 5, 19, 28, 30, 72). Several tested products successfully controlled wind erosion for a short time, but many were more expensive than equally effective wheat straw anchored with a rolling disk packer (30). The following are criteria for surface-soil stabilizers: (a) 100 percent of the soil surface must be covered, (b) the stabilizer must not adversely affect plant growth or emergence, (c) erosion must be prevented initially and reduced for at least two months, (d) the stabilizer should be easy to apply and not require special equipment, and (e) cost must be low enough for profitable use (5). Armbrust and Lyles (5) found five polymers and one resin-in-water emulsion that met all of these requirements. They added, however, that before soil stabilizers can be used on agricultural land, methods must be developed to apply large volumes rapidly. Also, reliable preemergent weed control chemicals for use on coarse-textured soils must be developed, as well as films that are resistant to raindrop impact while allowing water and plant roots to penetrate the soil, without adversely affecting the environment.

Barriers. Use of wind barriers is an effective method of reducing field width to control wind erosion (9). Hagen (51) and Skidmore and Hagen (93) developed a model that when used with local wind data shows wind barrier effectiveness in reducing wind erosion forces. Barriers will reduce wind erosion forces more than they will windspeed. A properly oriented barrier, when winds predominate from a single direction, will reduce wind erosion forces by more than 50 percent from the barrier leeward to 20 times its height; the reduction will be greater for shorter distances from the barrier.

Different combinations of trees, shrubs, tall-growing crops, and grasses can reduce wind erosion. Aside from conventional tree windbreaks (42, 84, 110), many other barrier systems are used to control wind erosion. They include annual crops, such as small grains, corn, sorghum, sudangrass, sunflowers (13, 47, 48, 52, 61); tall wheatgrass (1, 10); sugarcane; and rye strips on sands (John Griffin, SCS agronomist, Gainesville, Florida, personal communication, 1975).

Most barrier systems for controlling wind erosion, however, occupy space that could otherwise be used for crop production. Perennial barriers grow slowly and often are difficult to establish (36, 110). Such barriers also compete with crops for water and plant nutrients (74). As a result, the net effect of many tree-barrier systems is that their use may not benefit crop production (46, 79, 94, 95, 101).

Perhaps tree-barrier systems could be designed so that they become a useful crop, furnishing nuts, fruit, or wood.

Stripcropping. The practice of farming land in narrow strips on which the crop alternates with fallow is an effective aid in controlling wind erosion (*21*). Strips are most effective when they are at right angles to the prevailing wind direction, but they also provide some protection from winds that are not perpendicular to the field strip.

Stripcropping reduces erosion damage by reducing the distance the wind travels across exposed soil, localizing drifting that starts at a focal point, and reducing wind velocity across the strip when adjacent fields are covered with tall stubble or crops.

Although each method to control wind erosion has merit and application, establishing and maintaining vegetative cover, when feasible, remains the best defense against wind erosion. However, that becomes a difficult challenge as pressure increases to use crop residues for livestock feed and fuel for cooking.

Conclusions

Investigations of the factors influencing wind erosion led to the development of a wind erosion equation. The two-fold purpose of the wind erosion equation is to predict average annual soil loss from a field for specified conditions and to guide the design of wind erosion control practices.

Principles suggested by the wind erosion equation for controlling wind erosion include stabilizing erodible surface soil with various materials; producing a rough, cloddy surface; reducing field width or the distance wind travels in crossing a field unprotected with barriers and strips of crops; and establishing and maintaining sufficient vegetative cover. This last item is sometimes referred to as the "cardinal rule" for controlling wind erosion.

Although the wind erosion equation is extremely useful and widely applicable, users are cautioned that the value obtained for E is an estimate of average annual potential soil loss. The actual soil loss may differ from the potential because of (a) variation from the average of wind and precipitation, (b) inaccuracies in converting from relative field erodibility to annual soil losses, (c) relationships among variables not well defined for all combinations of field and climatic conditions, (d) seasonal variation of field erodibility, and (e) uncertainties inherent in the empiricism used in developing the equation.

Research in progress to improve the accuracy and applicability of the wind erosion equation includes:

► Determining the percentage of eroding soil that can be suspended during erosion under a wide range of field conditions and the residence time and fate of the various sizes of particles suspended by wind erosion.

► Refining the soil moisture term of the climatic factor, C, in the wind erosion equation. The current procedures assume that effective moisture of the surface soil particles varies with the PE index or dryness ratio, but surface moisture content is transient. Drying rate and dryness of particles, as a function of soil hydraulic properties and climatic variables, need to be examined and then related to the wind erosion process.

► Converting wind erosion prediction from a deterministic to a stochastic model by incorporating probability functions for some of the dynamic variables.

► Developing more applicable flux equations that can be integrated over time and space to predict soil erosion during single windstorms. Soil flux from fields that contain nonerodible elements decreases with time, which suggests that a time function is needed in the prediction equation.

REFERENCES

1. Aase, J. K., F. H. Siddoway, and A. L. Black. 1976. *Perennial grass barriers for wind erosion control, snow management, and crop production*. In *Shelterbelts on the Great Plains, Proceedings of the Symposium*. Publication No. 78. Great Plains Agricultural Council, Lincoln, Nebraska. pp. 69-78.
2. Anderson, C. H. 1975. *A history of soil erosion by wind in the Palliser Triangle of western Canada*. Historical Series No. 8. Research Branch, Canada Department of Agriculture, Ottawa, Ontario. 25 pp.
3. Apt, K. E. 1976. *Applicability of the Weibull distribution to atmospheric activity data*. Atmospheric Environment 10: 777-782.
4. Armbrust, D. V., and J. D. Dickerson. 1971. *Temporary wind erosion control: Cost and effectiveness of 34 commercial materials*. Journal of Soil and Water Conservation 26: 154-157.
5. Armbrust, D. V., and L. Lyles. 1975. *Soil stabilizers to control wind erosion*. Soil Conditioners 7: 77-82.
6. Armbrust, D. V., and L. Lyles. 1985. *Equivalent wind erosion protection from selected growing crops*. Agronomy Journal 77: 703-707.
7. Armbrust, D. V., W. S. Chepil, and F. H. Siddoway. 1964. *Effects of ridges*

on erosion of soil by wind. Soil Science Society of America Proceedings 28: 557-560.

8. Bagnold, R. A. 1943. <i>The physics of blown sand and desert dunes.</i> William Morrow & Co., New York, New York.
9. Bates, C. G. 1911. <i>Windbreaks: Their influence and value.</i> Bulletin 86. Forest Service, U.S. Department of Agriculture, Washington, D.C. 100 pp.
10. Black, A. L., and F. H. Siddoway. 1971. <i>Tall wheatgrass barriers for soil erosion control and water conservation.</i> Journal of Soil and Water Conservation 26: 107-110.
11. Black, A. L., and J. F. Power. 1965. <i>Effect of chemical and mechanical fallow methods on moisture storage, wheat yields, and soil erodibility.</i> Soil Science Society of America Proceedings 29: 465-468.
12. Budyko, M. I. 1958. <i>The heat balance of the earth's surface.</i> Nina A. Stepanova, Translator. U.S. Department of Commerce, Washington, D.C. 259 pp.
13. Carreker, J. R. 1966. <i>Wind erosion in the Southeast.</i> Journal of Soil and Water Conservation 11: 86-88.
14. Chepil, W. S. 1944. <i>Utilization of crop residues for wind erosion control.</i> Scientific Agriculture 24: 307-319.
15. Chepil, W. S. 1945. <i>Dynamics of wind erosion: III. The transport capacity of the wind.</i> Soil Science 60: 475-480.
16. Chepil, W. S. 1946. <i>Dynamics of wind erosion: V. Cumulative intensity of soil drifting across eroding fields.</i> Soil Science 61: 257-263.
17. Chepil, W. S. 1950. <i>Properties of soil which influence wind erosion: II. Dry aggregate structure as an index of erodibility.</i> Soil Science 69: 403-414.
18. Chepil, W. S. 1954. <i>Seasonal fluctuations in soil structure and erodibility of soil by wind.</i> Soil Science Society of America Proceedings 18: 13-16.
19. Chepil, W. S. 1955. <i>Effects of asphalt on some phases of soil structure and erodibility by wind.</i> Soil Science Society of America Proceedings 19: 125-128.
20. Chepil, W. S. 1956. <i>Influence of moisture on erodibility of soil by wind.</i> Soil Science Society of America Proceedings 20: 288-292.
21. Chepil, W. S. 1957. <i>Width of field strip to control wind erosion.</i> Technical Bulletin No. 92. Kansas Agricultural Experiment Station, Manhattan.
22. Chepil, W. S. 1958. <i>Soil conditions that influence wind erosion.</i> Technical Bulletin No. 1185. U.S. Department of Agriculture, Washington, D.C. 40 pp.
23. Chepil, W. S. 1959. <i>Wind erodibility of farm fields.</i> Journal of Soil and Water Conservation 14: 214-219.
24. Chepil, W. S. 1960. <i>Conversion of relative field erodibility to annual soil loss by wind.</i> Soil Science Society of America Proceedings 24: 143-145.
25. Chepil, W. S., F. H. Siddoway, and D. V. Armbrust. 1962. <i>Climatic factor for estimating wind erodibility of farm fields.</i> Journal of Soil and Water Conservation 17: 162-165.
26. Chepil, W. S., and N. P. Woodruff. 1954. <i>Estimations of wind erodibility of field surfaces.</i> Journal of Soil and Water Conservation 9: 257-265.
27. Chepil, W. S., and N. P. Woodruff. 1959. <i>Estimations of wind erodibility of farm fields.</i> Production Research Report No. 25. Agricultural Research Service, U.S. Department of Agriculture, Washington, D.C. 21 pp.
28. Chepil, W. S., and N. P. Woodruff. 1963. <i>The physics of wind erosion and its control.</i> Advances in Agronomy 15: 211-302.
29. Chepil, W. S., N. P. Woodruff, and A. W. Zingg. 1955. <i>Field study of wind erosion in western Texas.</i> SCS-TP-125. U.S. Department of Agriculture, Washington, D.C. 60 pp.

30. Chepil, W. S., N. P. Woodruff, F. H. Siddoway, D. W. Fryrear, and D. V. Armbrust. 1963. *Vegetative and nonvegetative materials to control wind and water erosion.* Soil Science Society of America Proceedings 27: 86-89.

31. Chepil, W. S., N. P. Woodruff, F. H. Siddoway, and L. Lyles. 1960. *Anchoring vegetative mulches.* Agricultural Engineering 41: 754-755, 759.

32. Chepil, W. S., and R. A. Milne. 1941. *Wind erosion of soils in relation to size and nature of the exposed area.* Scientific Agriculture 21: 479-487.

33. Claflin, L. E., D. L. Stuteville, and D. V. Armbrust. 1973. *Windblown soil in the epidemiology of bacterial leaf spot of alfalfa and common blight of beans.* Phytopathology 63: 1,417-1,419.

34. Cole G. W., L. Lyles, and L. J. Hagen. 1983. *A simulation model of daily wind erosion soil loss.* Transactions, American Society of Agricultural Engineers. 26: 1,758-1,765.

35. Daniel, H. A., and W. H. Langham. 1936. *The effect of wind erosion and cultivation on the total nitrogen and organic matter content of soils in the Southern High Plains.* Journal of American Society of Agronomy 28: 587-596.

36. Dickerson, J. D., N. P. Woodruff, and E. E. Banbury. 1976. *Techniques for improving survival and growth of trees in semiarid areas.* Journal of Soil and Water Conservation 1: 63-66.

37. Dregne, H. E. 1976. *Soils of the arid regions.* Elsevier Scientific Publishers Co., New York, New York. 237 pp.

38. Dudley, F. L. 1959. *Progress of research on stubble mulching in the Great Plains.* Journal of Soil and Water Conservation 14: 7-11.

39. Englehorn, C. L., A. W. Zingg, and N. P. Woodruff. 1952. *The effects of plant residue cover and clod structure on soil losses by wind.* Soil Science Society of America Proceedings 16: 29-33.

40. Erickson, D. A., P. C. Deutsch, D. L. Anderson, and T. A. Sweeney. 1984. *Wind driven interactive wind erosion estimator.* Agronomy Abstracts 247.

41. Fenster, C. R., G. A. Wicks, and D. E. Smika. 1973. *The role of eco-fallow for reducing energy requirements for crop production.* Agronomy Abstracts 122.

42. Ferber, A. E. 1969. *Windbreaks for conservation.* Agriculture Information Bulletin 339. Soil Conservation Service, U.S. Department of Agriculture, Washington, D.C.

43. Fisher, P. S., and E. L. Skidmore. 1970. *WEROS: A Fortran IV program to solve the wind erosion equation.* ARS 41-174. Agricultural Research Service, U.S. Department of Agriculture, Washington, D.C. 13 pp.

44. Food and Agriculture Organization, United Nations. 1960. *Soil erosion by wind and measures for its control on agricultural lands.* Development Paper No. 71. Rome, Italy.

45. Food and Agriculture Organization, United Nations. 1979. *A provisional methodology for soil degradation assessment.* Rome, Italy.

46. Frank, A. B., D. G. Harris, and W. O. Willis. 1977. *Growth and yields of spring wheat as influenced by shelter and soil water.* Agronomy Journal 69: 903-906.

47. Fryrear, D. W. 1963. *Annual crops as wind barriers.* Transactions, American Society of Agricultural Engineers. 6: 340-342, 352.

48. Fryrear, D. W. 1969. *Reducing wind erosion in the Southern Great Plains.* MP-929. Texas A&M University, College Station.

49. Gillette, D. A., I. H. Blifford, Jr., and C. R. Fenster. 1972. *Measurements of aerosol size distribution and vertical fluxes of aerosols on land subject to wind erosion.* Journal of Meteorology 11: 977-987.

50. Gore, R. 1979. *The desert: An age-old challenge grows.* National Geographic 156: 594-639.
51. Hagen, L. J. 1976. *Windbreak design for optimum wind erosion control.* In *Shelterbelts on the Great Plains, Proceedings of the Symposium.* Publication No. 78. Great Plains Agricultural Council, Lincoln, Nebraska. pp. 31-36.
52. Hagen, L. J., E. L. Skidmore, and J. D. Dickerson. 1972. *Designing narrow strip barrier systems to control wind erosion.* Journal of Soil and Water Conservation 27: 269-270.
53. Hagen, L. J., L. Lyles, and E. L. Skidmore. 1980. *Application of wind energy to Great Plains irrigation pumping. Appendix B: Method of determination and summary of Weibull parameters at selected Great Plains weather stations.* ATT-NC-4. Science and Education Administration, U.S. Department of Agriculture, Washington, D.C. 20 pp.
54. Hagen, L. J., and N. P. Woodruff. 1973. *Air pollution from dust storms in the Great Plains.* Atmospheric Environment 7: 323-332.
55. Halsey, C. F., W. F. Detmer, L. A. Cable, and E. C. Ampe. 1983. *SOILEROS—a friendly erosion estimation program for the personal computer.* Agronomy Abstracts 20.
56. Hare, F. K. 1983. *Climate and desertification A revised analysis.* Report No. 44. World Climate Programme, Geneva, Switzerland.
57. Harris, R. F., G. Chesters, and O. N. Allen. 1966. *Dynamics of soil aggregation.* Advances in Agronomy 18: 107-169.
58. Hayes, W. A. 1965. *Wind erosion equation useful in designing northeastern crop protection.* Journal of Soil and Water Conservation 20: 153-155.
59. Hayes, W. A. 1966. *Guide for wind erosion control in the northeastern states.* Soil Conservation Service, U.S. Department of Agriculture, Washington, D.C.
60. Hayes, W. A. 1972. *Designing wind erosion control systems in the Midwest Region.* RTSC-Technical Note, Agronomy L1-9. Soil Conservation Service, U.S. Department of Agriculture, Washington, D.C.
61. Hoag, B. K., and G. N. Geiszler. 1971. *Sunflower rows to protect fallow from wind erosion.* North Dakota Farm Research 28: 7-12.
62. Hurt, R. D. 1981. *The dust bowl: An agricultural and social history.* Nelson-Hall, Chicago, Illinois. 214 pp.
63. Jaenicke, R. 1979. *Monitoring and critical review of the estimated source strength of mineral dust from the Sahara.* In Christer Morales [editor] *Saharan Dust. Mobilization, Transport, Deposition.* John Wiley & Sons, New York, New York. pp. 233-242.
64. Johnson, G. L. 1978. *Economic design of wind electric systems.* Institute of Electrical and Electronic Engineers Transactions, Power Apparatus Systems PAS-97(2): 554-562.
65. Johnson, V. 1947. *Heaven's tableland: The dust bowl story.* Farrar-Straus, New York, New York. pp. 155-157.
66. Justus, C. G., W. R. Hargraves, and A. Mikhail. 1976. *Reference windspeed distributions and height profiles for wind turbine design and performance evaluation applications.* Technical Report, Contract No. E(40-1)-5108. Georgia Institute of Technology, Atlanta.
67. Lyles, L. 1974. *Speculation on the effect of wind erosion on productivity.* Special Report to Task Force on Wind Erosion Damage Estimates. U.S. Department of Agriculture, Washington, D.C.
68. Lyles, L. 1975. *Possible effects of wind erosion on soil productivity.* Journal of Soil and Water Conservation 30: 279-283.

69. Lyles, L. 1983. *Erosive wind energy distributions and climatic factors for the West.* Journal of Soil and Water Conservation 38: 106-109.

70. Lyles, L., and B. E. Allison. 1980. *Range grasses and their small grain equivalents for wind erosion control.* Journal of Range Management 33: 143-146.

71. Lyles, L., and B. E. Allison. 1981. *Equivalent wind-erosion protection from selected crop residues.* Transactions, American Society of Agricultural Engineers. 24(2): 405-408.

72. Lyles, L., D. V. Armbrust, J. D. Dickerson, and N. P. Woodruff. 1969. *Spray-on adhesives for temporary wind erosion control.* Journal of Soil and Water Conservation 24: 190-193.

73. Lyles, L., and J. Tatarko. 1982. *Emergency tillage to control wind erosion: Influences on winter wheat yields.* Journal of Soil and Water Conservation 37: 344-347.

74. Lyles, L., J. Tatarko, and J. D. Dickerson. 1984. *Windbreak effects on soil water and wheat yield,* Transactions, American Society of Agricultural Engineers. 27: 69-72.

75. Lyles, L., N. F. Schmeidler, and N. P. Woodruff. 1973. *Stubble requirements in field strips to trap windblown soil.* Research Publication 164. Kansas Agricultural Experiment Station, Manhattan. 22 pp.

76. Malin, J. C. 1946. *Dust storms—part one, two, and three. 1850-1860, 1861-1880, 1881-1900, respectively.* The Kansas Historical Quarterly 14: 129-144, 265-296, 391-413.

77. Mannering, J. V., and C. R. Fenster. 1983. *What is conservation tillage?* Journal of Soil and Water Conservation 38: 141-143.

78. McCalla, T. M., and T. J. Army. 1961. *Stubble mulch farming.* Advances in Agronomy 13: 125-196.

79. McMartin, W., A. B. Frank, and R. H. Heintz. 1974. *Economics of shelterbelt influence on wheat yields in North Dakota.* Journal of Soil and Water Conservation 29: 87-91.

80. Moldenhauer, W. C., and E. R. Duncan. 1969. *Principles and methods of wind-erosion control in Iowa.* Special Report No. 62. Iowa State University, Ames.

81. Olmstead, L. B. 1946. *The effect of long-time cropping systems and tillage practices upon soil aggregation at Hays, Kansas.* Soil Science Society of America Proceedings 11: 89-92.

82. Page, J. B., and C. J. Willard. 1946. *Cropping systems and soil properties.* Soil Science Society of America Proceedings 11: 81-88.

83. PEDCO-Environmental Specialists, Inc. 1973. *Investigations of fugitive dust-sources, emissions, and control.* Report prepared under Contract No. 68-02-0044, Task Order No. 9. U.S. Environmental Protection Agency, Office of Air Quality Planning and Standards, Cincinnati, Ohio.

84. Read, R. A. 1964. *Tree windbreaks for the Central Great Plains.* Agriculture Handbook 250. U.S. Department of Agriculture, Washington, D.C.

85. Secretariat of the United Nations Conference on Desertification. 1977. *Desertification: Its causes and consequences.* Pergamon Press, New York, New York.

86. Siddoway, F. H., W. S. Chepil, and D. V. Armbrust. 1965. *Effect of kind, amount, and placement of residue on wind erosion control.* Transactions, American Society of Agricultural Engineers. 8: 327-331.

87. Skidmore, E. L. 1965. *Assessing wind erosion forces: Directions and relative magnitudes.* Soil Science Society of America Proceedings 29: 587-590.

88. Skidmore, E. L. 1982. *Soil and water management and conservation: Wind erosion.* In Victor J. Kilmer [editor] *Handbook of Soils and Climate in Agriculture.* CRC Press, Boca Raton, Florida. pp. 371-399.

89. Skidmore, E. L. 1983. *Wind erosion calculator: Revision of residue table.* Journal of Soil and Water Conservation 38: 110-112.

90. Skidmore, E. L. 1986. *Wind-erosion climatic erosivity.* Climate Change 9:195-208.

91. Skidmore, E. L., and F. H. Siddoway. 1978. *Crop residue requirements to control wind erosion.* In W. R. Oschwald [editor] *Crop Residue Management Systems.* Special Publication No. 31. American Society of Agronomy, Madison, Wisconsin. pp. 17-33.

92. Skidmore, E. L., J. B. Layton, D. V. Armbrust, and M. L. Hooker. 1986. *Soil physical properties as influenced by residue management.* Soil Science Society of America Journal 50(2): 415-419.

93. Skidmore, E. L., and L. J. Hagen. 1977. *Reducing wind erosion with barriers.* Transactions, American Society of Agricultural Engineers. 20: 911-915.

94. Skidmore, E. L., L. J. Hagen, D. G. Naylor, and I. D. Teare. 1974. *Winter wheat response to barrier-induced microclimate.* Agronomy Journal 66: 501-505.

95. Skidmore, E. L., L. J. Hagen, and I. D. Teare. 1975. *Wind barriers most beneficial at intermediate stress.* Crop Science 15: 443-445.

96. Skidmore, E. L., M. Kumar, and W. E. Larson. 1979. *Crop residue management for wind erosion control in the Great Plains.* Journal of Soil and Water Conservation 34: 90-96.

97. Skidmore, E. L., N. L. Nossaman, and N. P. Woodruff. 1966. *Wind erosion as influenced by row spacing, row direction, and grain sorghum population.* Soil Science Society of America Proceedings 30: 505-509.

98. Skidmore, E. L., and N. P. Woodruff. 1968. *Wind erosion forces in the United States and their use in predicting soil loss.* Agriculture Handbook No. 346. U.S. Department of Agriculture, Washington, D.C.

99. Skidmore, E. L., P. S. Fisher, and N. P. Woodruff. 1970. *Wind erosion equation: Computer solution and application.* Soil Science Society of America Proceedings 34: 931-935.

100. Skidmore, E. L., W. A. Carstenson, and E. E. Banbury. 1975. *Soil changes resulting from cropping.* Soil Science Society of America Proceedings 39: 964-967.

101. Staple, W. J., and J. H. Lehane. 1955. *The influence of field shelterbelts on wind velocity, evaporation, soil moisture, and crop yields.* Canadian Journal of Agricultural Science 35: 440-453.

102. Svobida, L. 1940. *An empire of dust.* Caxton Printers, Ltd., Caldwell, Idaho. 203 pp.

103. Thornthwaite, C. W. 1931. *Climates of North America according to a new classification.* Geographical Review 25: 633-655.

104. U.S. Department of Agriculture, Soil Conservation Service. 1984. *National Resources Inventory.* Washington, D.C.

105. U.S. Department of Agriculture, Soil Conservation Service. 1986. *Wind erosion handbook* (draft). Washington, D.C.

106. Williams, J. R., C. A. Jones, and P. T. Dyke. 1984. *A modeling approach to determining the relationship between erosion and soil productivity.* Transactions, American Society of Agricultural Engineers. 27: 129-144.

107. Wilson, L. 1975. *Application of the wind erosion equation in air pollution surveys.* Journal of Soil and Water Conservation 30: 215-219.

108. Woodruff, N. P., and D. V. Armbrust. 1968. *A monthly climatic factor for the wind erosion equation.* Journal of Soil and Water Conservation 23: 103-104.

109. Woodruff, N. P., and F. H. Siddoway. 1965. *A wind erosion equation.* Soil Science Society of America Proceedings 29: 602-608.
110. Woodruff, N. P., J. D. Dickerson, E. E. Banbury, A. B. Erhart, and M. C. Lundquist. 1976. *Selected trees and shrubs evaluated for single-row windbreaks in the Central Great Plains.* NC-37. Agricultural Research Service, U.S. Department of Agriculture, Washington, D.C.
111. Woodruff, N. P., L. Lyles, F. H. Siddoway, and D. W. Fryrear. 1972. *How to control wind erosion.* Agriculture Information Bulletin No. 354. Agricultural Research Service, U.S. Department of Agriculture, Washington, D.C. 22 pp.
112. Woodruff, N. P., W. S. Chepil, and R. D. Lynch. 1957. *Emergency chiseling to control wind erosion.* Technical Bulletin 90. Kansas Agricultural Experiment Station, Manhattan.
113. Zingg, A. W. 1951. *A portable wind tunnel and dust collector developed to evaluate the erodibility of field surfaces.* Agronomy Journal 43: 189-191.
114. Zingg, A. W. 1951. *Evaluation of the erodibility of field surfaces with a portable wind tunnel.* Soil Science Society of America Proceedings 15: 11-17.
115. Zingg, A. W. 1953. *Wind-tunnel studies of the movement of sedimentary materials.* In Proceedings, Fifth Hydraulic Conference, Iowa Institute of Hydraulic, Research Bulletin 34. John Wiley & Sons, New York, New York. pp. 111-135.
116. Zingg, A. W., and N. P. Woodruff. 1951. *Calibration of a portable wind tunnel for the simple determination of roughness and drag on field surfaces.* Agronomy Journal 43: 191-193.

Appendix

Common units of measurements used in erosion research and conversion factors.

Conventional/ English Units	Metric Equivalents
Length:	
inch (in)	2.54 centimeters (cm)
foot (ft)	30.48 centimeters
mile (mi)	1.61 kilometers (km)
nauticle mile	1.85 kilometers
angotron (A°)	10^{-10} meters (m)
light-year	9.46×10^{-15} m
Energy, force, work, power and pressure	
pound (lb)	4.45 Newtons (N)
dyne	10^{-5} Newton
foot-pound	1.36 Joules (J)
British Thermal Unit (Btu)	252 calories (cal) or 1,054 J
Watt (W)	1 Joule/sec (J/S)
horsepower (hp)	746 Joules/sec
erg	10^{-7} Joule
atmosphere (atm)	1.013 bar = 1.013×10^5 Pascals (N/m²)
pound/inch² (lb/in²)	6.90×10^3 N/m²
Pascal (Pa)	1 Newton/m² (N/m²)
Weight, time, speed and angle	
day (d)	8.64×10^4 seconds (s)
year (yr)	3.156×10^7 seconds (s)
mile/hour (mi/h)	0.447 meter/second (m/s)
knot	0.5144 meter/second
degree (1°)	0.01745 radian (rad)
radian (rad)	57.30 degrees (°)
short ton	907 kilograms (kg)
pound (lb)	453.6 grams (g)
Volume	
cubic foot (ft³)	0.028 cubic meter (m³)
cubic inch (in³)	16.4 cubic centimeters (cm³)
gallon	3.785 liters (l)
gallon per acre	9.35 liters/hectare ($l \cdot ha^{-1}$)
quart (liquid)	0.946 liter (l)
Area and yield	
acre	4,047 square meters (m²)
acre	0.405 hectare (ha)
square foot (ft²)	9.29×10^{-2} square meter (m²)
square inch (in²)	645.2 square millimeters (mm²)
square mile (mi²)	2.59 square kilometers (km²)
pound per acre	1.12 kilograms per hectare ($kg \cdot ha^{-1}$)
pound per acre	1.12×10^{-3} megagrams per hectare ($Mg \cdot ha^{-1}$)

Conversion factors for units related to the universal soil loss equation (USLE).

To Convert From:	U.S. Customary Units	Multiply By:	To Obtain:	SI Units
Rainfall intensity, i or I	$\dfrac{\text{inch}}{\text{hour}}$	25.4	$\dfrac{\text{millimeter}}{\text{hour}}$	$\dfrac{\text{mm}}{\text{h}}$
Rainfall energy per unit of rainfall, e	$\dfrac{\text{foot-tonf}}{\text{acre-inch}}$	2.638×10^{-4}	$\dfrac{\text{megajoule}}{\text{hectare} \cdot \text{millimeter}}$	$\dfrac{\text{MJ}}{\text{ha} \cdot \text{mm}}$
Storm, energy, E	$\dfrac{\text{foot-tonf}}{\text{acre}}$	6.701×10^{-3}	$\dfrac{\text{megajoule}}{\text{hectare}}$	$\dfrac{\text{MJ}}{\text{ha}}$
Storm erosivity, EI	$\dfrac{\text{foot-tonf} \cdot \text{inch}}{\text{acre} \cdot \text{hour}}$	1.702×10^{-1}	$\dfrac{\text{megajoule} \cdot \text{millimeter}}{\text{hectare} \cdot \text{hour}}$	$\dfrac{\text{MJ} \cdot \text{mm}}{\text{ha} \cdot \text{h}}$
Storm erosivity, EI	$\dfrac{\text{hundreds of foot-tonf} \cdot \text{inch}}{\text{acre} \cdot \text{hour}}$*	17.02	$\dfrac{\text{megajoule} \cdot \text{millimeter}}{\text{hectare} \cdot \text{hour}}$	$\dfrac{\text{MJ} \cdot \text{mm}}{\text{ha} \cdot \text{h}}$
Annual erosivity, R†	$\dfrac{\text{hundreds of foot-tonf} \cdot \text{inch}}{\text{acre} \cdot \text{hour} \cdot \text{year}}$	17.02	$\dfrac{\text{megajoule} \cdot \text{millimeter}}{\text{hectare} \cdot \text{hour} \cdot \text{year}}$	$\dfrac{\text{MJ} \cdot \text{mm}}{\text{ha} \cdot \text{h} \cdot \text{y}}$
Soil erodibility, K‡	$\dfrac{\text{ton} \cdot \text{acre} \cdot \text{hour}}{\text{hundreds of acre} \cdot \text{foot-tonf} \cdot \text{inch}}$	1.317×10^{-1}	$\dfrac{\text{metric ton} \cdot \text{hectare} \cdot \text{hour}}{\text{hectare} \cdot \text{megajoule} \cdot \text{millimeter}}$	$\dfrac{\text{t} \cdot \text{ha} \cdot \text{h}}{\text{ha} \cdot \text{MJ} \cdot \text{mm}}$
Soil loss, A	$\dfrac{\text{ton}}{\text{acre}}$	2.242	$\dfrac{\text{metric ton}}{\text{hectare}}$	$\dfrac{\text{t}}{\text{ha}}$
Soil loss, A	$\dfrac{\text{ton}}{\text{acre}}$	2.242×10^{-1}	$\dfrac{\text{kilogram}}{\text{meter}^2}$	$\dfrac{\text{kg}}{\text{m}^2}$

*This notation, "hundreds of" means numerical values should be multiplied by 100 to obtain true numerical values in given units. For example, $R = 125$ (hundreds of ft-ton·in/acre·hr) = 12,500 ft-tonf·in/acre·hr. The converse is true for "hundreds of" in the denominator of a fraction.
†Erosivity, EI or R, can be converted from a value in U.S. customary units to a value in units of Newton/hour (N/hr) by multiplying by 1.702.
‡Soil erodibility, K, can be converted from a value in U.S. customary units to a value in units of metric ton·hectare/Newton·hour (t·h/ha·N) by multiplying by 1.317.

Index

Acacia species, 179
Acidity, 188, 195, 197
Acoustic devices
  depth measurement and, 58
  rainfall kinetic energy and, 150
  sediment monitoring and, 56
Aeration, 197
A factor, USLE, 15-17
Africa
  erosion indicators in, 179
  erosivity indexes and, 157
  river sediment yields, 2-4, 43
  SLEMSA, 102, 180-181
  See also specific countries
Aggregation
  cropping history and, 208
  erodibility indexes and, 142, 144-145
  erosion process model and, 127
  vegetative cover and, 166
  wind erodibility groups, 206-208
Agricultural Research Service, 102
Agroforestry, 182
Albedo, 188
Alfalfa, 208
Algeria, 43
Aluminum, 101, 188
Arid areas, 98, 203
Australia, 29, 136
Automatic pumping samplers, 46, 54-55

Barrier systems, for wind erosion,
  225-226

Boundary effects
  rainfall simulator plots, 86
  small erosion plots, 13-14
  unit-source watersheds, 32
Brazil, 166
Brown, Lester, 97
Brownell Creek, Nebraska, 60

California, 66
Carbon, sediment yield of, 10, 61
Centrifuges, 60
Centro International de Mejoramiento
  de Maiz y Trigo (CIMMYT), 181-182
C factor, USLE, 21-22, 28
  cropstages and, 180
  soil loss and, 133
  sub-factor method and, 102-103
Channel erosion
  CREAMS model and, 18
  river sediment yields and, 43
  USLE and, 17
Chemicals, Runoff, and Erosion for
  Agricultural Management Systems
  (CREAMS), 112-114
  EPIC and, 117
  sediment yield and, 104
  USLE factors in, 18
CIMMYT. *See* Centro International
  de Mejoramiento de Maiz y Trigo
Clay
  ratio, 144
  seasonal erodibility of, 206
  sediment yield and, 59

soil productivity and, 188
wind erodibility groups, 207
Colloid ratio, 142
Colorado
sediment yield and, 42, 60
wind erosion in, 205
Compaction, 188
Computers
CREAMS and, 113
erosion prediction and, 105
USLE and, 99
Concentrated flow erosion, 100, 104-105
Conductivity, of soil, 197-198
Contour rows
rill formation and, 11
USLE P factor and, 31
Coon Creek, Wisconsin, 66
Core analysis, lake sediment, 61-63
Corn. *See* Maize
Cotton
crop residue of, 218, 220
rainfall interception values, 176
vegetative cover and, 174
Cover. *See* Vegetative cover
Cowpea, 190, 193
Creedy River, England, 48-49, 52-53
Cropping systems
conservation research for, 181-183
erosion control and, 101-102
erosion prediction and, 179-180
P factor for, 29-31
rill formation and, 11
soil loss and, 167-168, 178
USLE evaluation of, 22
vegetative cover and, 169
wind erosion and, 223
Crop residue, wind erosion and,
216-224. *See also* Vegetative cover
Crop yield
agronomic measurement and, 189
desurfacing and, 193-194
EPIC model, 116-117
PI and, 114-116
predictive models, 106, 197-199
soil erosion and, 1, 99, 187-189
soil profile and, 190, 194-196
technological advances and, 97
tillage and, 178
wind erosion and, 208, 222

Dali River, People's Republic of
China, 40
Dart River, England, 48, 61
Densimeters, 56
Deposition, of sediment, 123-125
Desertification, 203
Desurfacing, 193-194
Detachment processes, 141-142
erosion process model and, 123-124
rainfall kinetic energy and, 150
runoff depth and, 156
time-dependence of, 4-5
vegetative cover and, 176
wind erosion and, 204
Devon, England, sediment yield data,
48-49, 52-53, 61
Dispersion ratio, 142, 144
Drainage
laboratory erosion plots, 14
sediment delivery ratio and, 64-65
Drifting, wind erosion and, 213
Driftless area, Wisconsin, 42
Dunes, 213
Dust, 222

Economic factors
sediment damage, 1-2
soil loss models, 106
vegetative cover, 168
Edge effects. *See* Boundary effects
Ellison splash cup method, 151-152
Enrichment ratio, 113, 194
Entrainment efficiency, 124, 128, 134
EPIC. *See* Erosion Productivity
Impact Calculator
Equal transit-rate sampling, 46
Erodibility, of soil, 141
crop residue and, 216
data reliability for, 4-5
dry soil fractions, 208
factors affecting, 101
field length and, 214-216
indexes of, 142
laboratory factors and, 144-145
nomogram estimation of, 147-149
plant indicators of, 179
rainfall simulations and, 90
seasons and, 206
SLEMSA and, 180

soil classification for, 141-142, 206
time-dependence of, 4-5
units of, 149
USLE factors for, 15-16, 19, 24-25, 143-147
wind erosion models and, 221-223
Erosion, definition of, 10-12. *See also* Soil erosion; specific types of erosion
Erosion Productivity Impact Calculator (EPIC), 106
climatic variability and, 116
CREAMS and, 117
wind erosion, 221
Erosion research plots
laboratory plots, 12, 99, 144
rainfall simulation and, 91-92
scale of, 10
small plots, 12-15
unit-source watersheds, 31-34
Erosivity, of rainfall, 141
data reliability for, 4-5
direct measurement of, 151-152
factors affecting, 101
indexes of, 16, 152-156
kinetic energy equations and, 23, 150
momentum and, 150
river basins, 157
sediment yield and, 104
single-storm events, 155
storm definition for, 25
time-dependence of, 4-5, 23
USLE R factor for, 15-17, 19-20, 157
Erosivity, of wind, 209-213. *See also* Wind erosion
Ethiopia, 43
Eucalyptus species, 166
Extrapolation
erosion prediction and, 99
sediment yield maps and, 2-4

Fallow plots
USLE erosion plots and, 18-19, 29
wind erosion and, 224
Fertilization, desurfacing and, 194
Floods, sediment yield sampling and, 51
Flumes, 32, 86
Fodder, as cover, 177
Forest cover, 166

Frains Lake, Michigan, 61-62

Gamma radiation, 56-57
Geologic erosion, 11. *See also* specific types of erosion
Georgia, 42
Grasslands
as cover, 177
erosion indicators for, 179, 219
Grazing, 203
Great Plains, United States, 203-204, 222
Greenhouse environments, 194
Gully erosion
forest cover and, 166
gully definition for, 100
prediction of, 104-105
river sediment and, 43

Hawaii
PI model and, 198
sediment loss in, 130
Henin erodibility index, 142, 144
Herbicides, wind erosion and, 224-225
Hocking River, Ohio, 60
Hong Kong, forest cover in, 166

Infiltration
laboratory erosion plots, 14
porosity and, 142
rainfall simulation and, 92
vegetative cover and, 165
Instability index, 142, 144
International Council for Research in Agroforestry, 182
Iron
sediment yield percentages, 61
soil erodibility and, 101
Irrigation
rainfall simulation and, 79
USLE and, 107

Java, 40

Kansas, 205, 209
Kentucky, 60
Kentucky rainfall simulator, 79, 81

Kenya
  agroforestry in, 182
  river sediment yields in, 40, 44-48
  Sagana River monitoring in, 57
K factor, USLE, 143-147
  fallow plot preparation for, 15-16, 143
  laboratory parameters for, 144-145
  storm definition for, 25
  time-dependence of, 24
Kinetic energy, of rainfall
  drop size and, 166
  erodibility and, 142, 145
  erosivity and, 23, 150-151
  simulated rainfall, 89
  USLE R factor and, 152-156
Kiowa Creek, Colorado, 60

Laboratory erosion plots, 12-15
  erodibility indexes, 144
  rainfall simulation and, 92
Lakes, sediment yield for, 60-63
L factor, USLE, 15-17, 20-21, 25-26
Loam
  entrainment efficiency for, 132
  seasonal erodibility of, 206
  wind erodibility groups and, 207
Loess, 132, 142
Lone Tree Creek, California, 66

Macrofauna, 188
Magnetic analysis, 63
Maize
  crop residue of, 218, 220
  crop rotation and, 208
  Nigerian crop yields, 190, 193
  USLE C factor and, 180
  wind erosion and, 225
Mali, 43
Manganese, 188
Michigan, lake sediment in, 61-63
Minnesota, soil PI for, 115
Mirror Lake, New Hampshire, 62
Modified Universal Soil Loss Equation
  (MUSLE), 17-18
Momentum, rainfall erosivity and,
  150
Mopane tree, 179
Mulch, 176
Multidivisor systems, 4

Mupundu, 179

Nebraska, 60
Nephelometers, 56
New Guinea, 40, 62
New Hampshire, 62
New Zealand, 40, 44
Nigeria
  crop yields in, 191-194
  erodibility indexes for, 149
  PI model and, 198
  rainfall kinetic energy in, 156
  river sediment yields, 43
Nitrogen
  erosion and, 188, 195
  maize and, 194
Nuclear probes, sediment measure-
  ment and, 46, 56-57, 67
Nutrients, 59, 188, 194

Ohio, 60
Oka River, Soviet Union, 66
Oklahoma, 45-46
Optical instrumentation, sediment
  transport and, 56
Orchards, as cover, 177
Organic matter
  erosion and, 195
  sediment yields and, 59, 61
  sufficiency of, 198
Overflow, 4

Palouse Region, United States, 103
People's Republic of China
  loess erosion in, 142
  river sediment yields in, 40
  sediment sampling in, 52
Pests, 188
P factor, USLE, 29-31, 180
Phosphorus, 61, 188, 194
Photography, vegetative cover
  methods and, 170
Plinthite, 188
Plowing
  soil erodibility and, 143
  USLE plots and, 20
Plum Creek, Kentucky, 60
Poland, 39
Pollution, sediment transport and, 59

Porosity, 142, 195
Potassium, 188, 194
Precipitation. *See* Rainfall
Productivity Index (PI), 106, 114-116
  definition of, 197
  Minnesota soils, 115
  See also Crop yield
Purdue infiltrometer, 79

Quadrat sighting, 170-172

Rainfall
  climatic factor and, 209-210
  crop hazard ratings, 174
  droplet size and, 150-151, 166
  erosivity of. *See* Erosivity, of
    rainfall
  kinetic energy of, 89, 100, 153
  rill erosion and, 11
  river sediment yield monitoring
    and, 54
  sediment transport model and,
    122-123
  single-storm events, 103
  SLEMSA and, 180
  splash from, 13-14
  storm definition for, 25
  unit-source watersheds, 32
  USLE R factor for, 15-17, 23,
    152-157
  vegetative cover and, 163-164, 175
  weed control and, 225
Rainfall simulation
  advantages and limitations of, 75-76
  cold weather and, 22
  crop yield research and, 192
  data interpretation, 93
  erodibility indexes and, 145-146
  erosion prediction and, 107
  laboratory erosion plots, 12-13
  natural rainfall and, 88-89
  raindrop sizes and, 77
  rainulator, 27, 79-80, 145
  requirements of, 85-86
  research procedures for, 87-93
  rilling and, 83
  scaling of, 10
  types of, 78-84
  vegetative cover and, 167

Rainulator, 79-80
  erodibility indexes and, 145
  USLE S factor and, 27
Rangeland, 203
Rape, 218
Rating curves, sediment yield, 47-48
Research projects, outlines for, 9-10
Reservoirs, sediment in, 58, 97
R factor, USLE, 15-17, 23, 152-157
  See also Erosivity, of rainfall
Ridge roughness, 213
Rill processes
  cropping systems and, 11
  erosion research plots and, 13
  rainfall simulation and, 83
  rill definition for, 100
  soil loss models and, 121-122
  stream power and, 130
  USLE and, 18, 102
River basins
  erosivity indexes, 157
  sediment yields, 1, 39-44
Root growth, 198
Rotation, of crops
  aggregation and, 208
  rainfall simulation and, 90
  soil loss and, 167-168
Row-cropping, rill formation and, 11
Runoff
  barriers to, 130
  collection system for, 192
  cover relation equation, 137
  depth versus soil detachment, 156
  erosivity of, 101, 104
  laboratory erosion plots, 14
  measurement of, 32-33
  rainfall simulation and, 86
  stemflow and, 168
  transport capacity of, 10
  USLE measurement, 17
  vegetative cover and, 164
  See also Sediment transport

Sagana River, Kenya, 57
Sahara desert, 204
Salinity, 188
Sand, 59, 142
Saturation, laboratory erosion plots, 14
Scotland, 62

Sediment delivery ration (SDR)
  basin area and, 64-65
  upstream erosion and, 42
  USLE and, 18, 103-104
  watershed area and, 101
  See also Runoff
Sediment transport
  analytic model for, 120-122
  downstream damage from, 97
  deposition model, 125
  erosion process model and, 123-130
  nutrients and, 59
  planar model, 121
  stream power and, 124-125
Sediment yield
  African rivers and, 2-4, 43
  automatic sampling of, 54-60
  channel erosion and, 43
  CREAMS and, 113
  data reliability for, 4-5, 44-48
  definition of, 10
  deposition and, 100
  global patterns, 1, 39-41
  human activity and, 63
  indirect calculation of, 47-48
  lake studies, 60-63
  land slope and, 135
  manual sampling of, 51-53
  monitoring programs for, 49-63
  off-site damage from, 1
  particle size composition, 60
  planar erosion model and, 127
  prediction models for, 103-104
  reservoir surveys and, 58
  sampling techniques, 45-46
  sediment properties, 59
  soil loss measurements and, 42
  South America, 2-3
  unit-source watersheds, 33
  upstream soil loss and, 63-67
  USLE estimation of, 18
  U.S.-series samplers for, 50
  water surface tension and, 11
  See also Sediment delivery ratio;
    Sediment transport
Seedbed preparation, flow erosion
  and, 100
S factor, USLE, 20-21, 26-28
Shade, 166

Sheet erosion, 99-100, 102, 166
Silica, erodibility and, 144
Silt
  erodibility nomogram and, 148
  erosion susceptibility of, 142
  sediment properties, 59
SLEMSA. *See* Soil Loss Estimator for
  Southern Africa
Slope factors
  contour banks and, 132
  erosion rate and, 101
  sediment yield and, 135
  soil loss and, 130-133
  USLE factor for, 15-17, 20-21,
    25-26, 146-147
  USLE rill systems and, 18
  vegetative cover and, 167
Sodium, soil erodibility and, 101
Soil, characteristics of
  crop yield and, 188, 190
  erosion influence on, 195
  thawing and, 100, 103
  tillage methods and, 169
  vegetative cover and, 163-164
Soil conservation
  farming systems research and, 180-182
  process approach to, 120
  USLE and, 17
  vegetative cover and, 163-164
  wind erosion control, 222
Soil erosion
  climatic factor for, 209
  contour banks and, 132
  cropping systems and, 167-168,
    177-178
  crop yield, and, 187-189
  desurfacing and, 193-194
  downstream sediment yields and,
    63-67
  economic models, 106
  erosion factor interdependence, 167
  forest cover and, 166
  geomorphological approach to, 196
  inter-rill erosion, 12
  land slope and, 136
  organic matter and, 163
  plot length and, 130-132
  prediction models for, 101-108
  process model for, 122-130

sediment transport model, 125
sediment yield measurement and, 42
types of, 99-100
vegetative indicators of, 178-179
See also Erodibility, of soil; Soil loss tolerances; Universal soil loss equation; specific types of erosion
Soil Loss Estimator for Southern Africa (SLEMSA), 102, 180
Soil loss tolerance (T factor)
    Australian soils, 136
    cover effects and, 133
    crop yields and, 196
    erosion prediction and, 98
    PI model and, 114-116
    slope and, 133
    weathering rate and, 196
Sonic depth recorders, 58
Sorghum
    crop residues of, 218, 220
    rainfall interception of, 174
    row spacing for, 178
    wind erosion and, 225
Soybeans, 218
Splash
    erodibility indexes and, 142
    erosivity measurement and, 151-152
    particle detachment and, 123-124
    small erosion plots, 13-14
    vegetative cover and, 176
Sprinklers
    rainfall simulation and, 79
    USLE and, 107
Steepness, USLE S factor for, 15-17, 20-21, 26-28
Stemflow runoff, 168
Storms
    definition of, 25
    erosivity index for, 155
    rainfall simulation of intensity, 87
    single-storm events, 103
    soil erosion models and, 99
Stream power, 124-125, 129-130
Stripcropping, 226
Sugarcane, 130, 225
Sulfur, 188
Sunflowers, 218, 225
Surface tension, runoff concentrations and, 11

Sweden, 62

Taiwan, 29, 40
Tana River, Kenya, 44-48
Tanzania, 149, 178
Terraces
    rill formation and, 11
    unit-source watersheds and, 31
T factor. *See* Soil loss tolerance
Thawing soil, 100, 103
Thornthwaite index, 209
Tillage
    crop yields and, 189
    erodibility and, 101
    flow erosion and, 105
    rainfall simulation and, 90
    rill erosion and, 100
    soil damage from, 169, 178
    USLE factors and, 31
    vegetative cover and, 166
    wind erosion and, 223
Tolerance, soil loss. *See* Soil loss tolerance
Trace elements, 188
Trees
    as cover, 166
    as erosion indicators, 178-179
    wind erosion and, 225
Tropical areas
    USLE and, 143, 180
    vegetative cover in, 163
Trummen Lake, Sweden, 62
Turbidity meter, sediment yield measurement and, 46, 48, 56-57

Ultrasonic sensors, 56, 150
United Nations Environmental Programme, 1
United States Water Resources Council, 51
Universal soil loss equation (USLE), 99, 111-112
    applicability of, 119-120
    cropping system evaluation and, 23
    erosion plots for, 15-31, 143
    factors for, 15-16, 20-21, 102, 143-147
    fallow plots for, 29-30
    geographical applicability of, 102
    limitations of, 107

runoff measurement for, 17
sediment yield and, 18, 103
single-storm events, 103
vegetative cover and, 21-22,
    102-103, 179-180
See also Soil erosion; specific factors

Vegetative cover
ambivalent effects of, 166-167
as erosion indicator, 178-179
canopy quality and, 160
cotton and, 174, 176, 218
crop classification and, 177
cropping systems and, 169
economic aspects of, 168
erodibility and, 101
fodder as, 177
forest cover, 166, 182
measurement methods, 170-176
rainfall and, 163-164, 175
runoff relation equation, 137
soil and, 164-165, 133-137
soil cover rating, 174
USLE C factor and, 21-22, 102-103
wind erosion and, 203, 216-224
Versitols, 132, 134

Washita River, Oklahoma, 45-46
Water
depth of, 58
for rainfall simulators, 85
surface tension of, 11
See also Rainfall; Runoff

Watersheds, unit-source research
    plots, 31-34
Weathering rate, 196
Weeds
USLE factors and, 22
vegetative cover and, 166
wind erosion and, 225
Wetting, heat of, 145
Wheat, crop residues of, 218
Wibull distribution, 211-212
Wind erodibility groups (WEGs), 207
Wind erosion
barriers and, 225-226
climatic factors for, 213
erodibility and, 204-208
erosivity and, 209-213
field length and, 214-215
model for, 221-223
ridge roughness and, 213
soil stabilizers and, 224
susceptibility to, 203
Wisconsin, sediment yield data, 66
Wollny, Ewold, 164
Worldwatch Institute, 97

Yangtze River, People's Republic of
    China, 52
Yellow River, People's Republic of
    China, 40, 142

Zimbabwe, 156, 174
Zinc, 188

Photo Credits

Page xiv   Food and Agriculture Organization, United Nations
Page 38   R. E. Dils
Page 74   Agricultural Research Service, L. D. Meyer
Page 96   Agricultural Research Service, L. D. Meyer
Page 140   U. S. Department of Agriculture
Page 162   M. A. Stocking
Page 202   Soil Conservation Service, Gene Alexander